Recollections of East Fife Fisher Folk

Recollections of East Fife Fisher Folk

Belle Patrick

Preface by
Patrick Collinson

First published in 2003 by
Birlinn Limited
West Newington House
10 Newington Road
Edinburgh
EH9 1QS

www.birlinn.co.uk

ISBN 1 84158 281 6

British Library Cataloguing-in-Publication Data
A catalogue record of this book is available from the British Library

Typset by Textype, Cambridge
Printed and bound by Cox & Wyman Ltd, Reading

Contents

Preface

About Belle Patrick

THE AUTHOR OF THIS BOOK, my mother, was born Belle Hay Patrick on 22 August 1895. My (only) Christian name is my mother's surname. Whether either of us was entitled to the name 'Patrick' became an interesting question when my mother's mother died, in her nineties, in 1956. It was well enough known that her husband, known as Alec Patrick, was illegitimate. Born in West Fife to a fishing family but sent at a tender age into the coalmines, he had arrived on the scene when his father had been away at sea for some time. There is even a family tradition that the true father was the earl of Wemyss, who owned much of that coal. But it had always been assumed that his mother had registered her son in her husband's name, Patrick. Only in 1956 was the certificate uncovered which revealed that she had done no such thing, but had brazenly registered Alec in her own name, to which, as a Scot, she was certainly entitled. It appears that insofar as the Patricks have brains, they derive from my great-grandmother, who was known to be a clever woman, if far from 'guid livin'.

As Belle Patrick explains, her father got himself out of the pits at the age of sixteen and walked eastwards along the coast of the Firth of Forth, looking for employment. He struck lucky when he got to Anstruther and found an opening as an apprentice fish-buyer. Belle Patrick's maternal grandfather, whose name was Pringle, was clerk to the Harbour Commissioners of Anstruther. Belle tells the story of how his father, not one of the fisherfolk, lived in that other world, inland, farming Fife, but not far enough from the coast to escape being pressganged in the Napoleonic Wars. He died in a French prison, leaving a wife and his many children destitute, a strange prolepsis of what would happen to his granddaughter, a hundred years later. Belle tells how, while her grandfather looked after the kine, the son of the laird became a friend and came back from school day by day to teach him to read, in the fields. He later became a poet, most of his verses being written in the cause of temperance. What took Grandfather Pringle to the coast is not known, but once there he rose to his position in the little hierarchy of Anstruther harbour management. So Anstruther harbour brought Agnes Pringle and Alec Patrick together. Agnes was cheeky enough, and ran after red-haired Alec shouting, 'Carrots!' But then Alec had a serious accident, and Agnes, full of remorse, visited him in hospital. It was not long before they married, I suppose in the later 1870s.

Belle was the eighth of ten children. There was Jim, of whom I know little, and then Dave, whom I knew well.

Dave's father had tried to put him behind the counter of a shop, but he was only interested in boats and fishing, and he ran away to the army, not to be seen again by the family for many years. What stories he had to tell! And he knew just where the biggest fish were to be caught, and at what stage of the tide. Then came three girls – Lizzie, Chrissie, Nettie, who all married and had families, Chrissie in Canada. There was perhaps a small gap before Annie, and then Willie, who also made his life in Canada, Belle (or Bella, it was never quite clear), Rita, who never married but looked after her mother into her great old age, and John. If my granny wanted little John the only way she could get to his name was to say, 'Annie–Willie–Bella–Rita–**John**!'

Belle Patrick reports of her father that he was of the 'Dippie' persuasion, a Baptist who was induced to take himself and his family off to the 'auld kirk' by the offer of a stipend as precentor which equalled the annual rent of his house close by Anstruther harbour. She also tells us that he was saved from a drink problem by a religious conversion, and that he preached and sang in the open air, especially to the Tinkers, who used to put up their tents and caravans in the lanes between Cellardyke and Kilrenny. She does not say that the drink took over again, although I suspect that it did. Certainly Alec suffered from depressive symptoms. In the autumn of 1904 he was on his way, as a fish-buyer, to the annual herring fishing off the East Anglian coast when he was taken ill in Newcastle. A brain tumour was diagnosed, and Alec decided that he could now be only a burden to his large family. A few days later, in Lowestoft, he told his landlady that he was going for a walk, and he was not seen alive again. Fish-buyers as a matter of routine carried revolvers to protect the considerable sums of cash in their charge, and he had shot himself. It was six weeks before his body was found, in a ditch. On that night my granny went to her weekly prayer meeting and heard voices: 'Shall we tell her now or later?' My Auntie Rita, lying in bed with the other children, recalled the commotion when everyone came back to the house. Ten-year old Willie said, 'That's my feyther deid, and I'll need to get a job.'

Willie's job in the butcher's shop brought some meat home at weekends. Fishermen in the harbour knew about the plight of the Patricks and always sent the children home with fish. Otherwise there was oatmeal. For the rest of her days Auntie Rita could never abide the smell of fish, which was a problem since my childhood holidays in Granny's house were entirely occupied with fishing.

Granny's brother Tom ('oor Tam') had emigrated to South Africa, where he had prospered. He sent the money to purchase a smart little semi-detached villa in the genteel world of West Anstruther, which, in honour of South Africa, was called 'Algoa', the address Shore Road, a few steps from the Billowness. There my granny spent a widowhood which lasted fifty-two years. She was a godly old lady who read her Bible to the last, a large-print edition which still required a magnifying glass to make it legible. 'Curses on Babylon! curses on Babylon! whit way should I strain my een reading curses on Babylon!', 'Yon Paul was an awfie man to argie!' When the raids on Berlin began in the war she cried out to Rita: 'It's the Lord's hand this time richt enough! There's moskeetees attacking Berlin!'

My mother shared in the all-pervasive seriousness and godliness which came from the Pringle side of the family. In a sense she was born out of one of Anstruther's religious revivals, since her second name of Hay (which we have passed on to our elder daughter) was the name of one of the Faith Mission 'pilgrims' who was lodging with the Patricks at the time of her birth. As a child she learnt the entire Book of Psalms by heart and never forgot them. The last of the revivals happened in 1920, a grass fire which began among the fishermen themselves, and out of this came the Children's Meeting which Belle organised for as many as 200 children, and which she describes in the Epilogue to this book. And it was that which reintegrated her into the fishing community of Cellardyke and supplied the knowledge which the book exploits. Wherever she went in the town she was surrounded and followed by a cluster of bairns, and their mothers and grannies remembered that this was 'Alec Patrick's lassie'.

Belle had more than her fair share of the Patrick brains, although she was not, at first, very practical, in the domestic sense. Her mother sometimes called her 'a gey haundless tottie'. Not long after she was orphaned at nine years of age she won a scholarship to the local secondary school, the Waid Academy, and that led on to a bursary at St Andrews University. But her mother would not allow her to take up the place and insisted that she get a job. The job – shorthand and typing for the local firm of Mackintosh and Watson, solicitors and bank agents – at first paid less than the bursary, a mere £5 a year, rising to a dizzy £12.10s after four years. Belle started in May 1911. But in August 1915 she was moved to neighbouring Pittenweem as managing clerk to Mr Mackintosh, who was Town Clerk of the Burgh. There her salary eventually rose to £120 per annum. It was wartime, and to all intents and purposes Belle Patrick *was* the Town Clerk, responsible among many other things for collecting the rates and balancing the books, without any discrepancy, for the first time in living memory. When she applied for the position of Head Clerk to the Roxburgh Education Authority, Mackintosh's reference described her as 'possessed of exceptional parts'. But Mr Watson, the most kindly and appreciative of employers, knew that in the brave new world which was supposed to follow the war, women were likely to qualify as lawyers. So instead of pursuing an administrative career in Roxburghshire, in August 1920 Belle Patrick was formally apprenticed to Mackintosh and Watson as a trainee solicitor. She read her law books early in the morning. Five years later, in October 1925, she passed her legal examinations and was admitted a 'Law Agent (or Solicitor) to practise in all the Courts of Scotland, Supreme as well as Inferior'. I still have some of her examination papers, written in both shorthand and longhand and addressing knotty questions to do with trusts, health and safety, the obligations incurred in gambling transactions, conveyancing, and other things of that kind. My mother was one of the first woman lawyers in Scotland. Was she or that other Fifer Jennie Lee, the politician and wife of Aneurin Bevan, the very first?

But she never practised. Instead she responded to a call to become a kind of missionary in Algeria. In her words it was to be 'the Gospel instead of the Law'. I say 'a kind of missionary' since her role was that of secretary to the remarkable Miss Lilias Trotter, founder and leader of the Algiers Mission Band, on a salary of £30 a year. Miss Trotter had been a favourite pupil of John Ruskin, and when she left him for North Africa he had said 'What a waste!' But Miss Lilias Trotter continued to execute superb little watercolours and pastels of the Algerian scene, which were published in books with such titles as *Between the Desert and the Sea*. And she wrote devotional books, and many, many letters. Like other great Victorian ladies, she was confined to bed, and that was where my mother came in. She lived with Miss Trotter in Dar Naama, the rambling house of a former Barbary corsair in El Biar, an eastern suburb of Algiers. The house was full of lady missionaries, since the Algiers Mission Band consisted exclusively of ladies, their mission not just to the women of Algeria but to the entire population. There were also English *women* of a lower social class who looked after the housekeeping, like the lay brothers in a medieval monastery. No-one knew quite where to place my mother, who clearly came from the humblest of backgrounds but was highly educated.

The Algiers posting, initially for one year, had been negotiated even before it was known whether Belle had passed her exams and she was at Dar Naama by late November 1925. Lilias Trotter wrote one enthusiastic letter after another to Anstruther: 'That dear daughter of yours' – 'I do look on her as one of God's most special gifts to me of late.' My mother, guessing what was being said, wrote: 'Please remember she's English and like all of them here is apt to gush a bit.' Lilias Trotter explained to

Mrs Patrick that Belle's first task had been 'getting the Gospel to a clan of Moslems, numbering hundreds of thousands in this land alone, for whom as yet hardly anything has been done – those belonging to the brotherhood of mystics'. In other words, she was typing the text of one of Miss Trotter's books, aimed at the Sufis. But a year later Lilias Trotter told Belle that 'she never could quite see me as a missionary, at least not yet', and proposed that she go home, more precisely to Croydon, to become the link person for the various prayer and support groups linked to the Band which were springing up around the country. But first it was necessary that she should learn more about the mission by joining some of the 'tournées' undertaken in the interior of Algeria and Tunisia, the purpose of which was 'colportage', distributing Christian literature.

Belle Patrick had some hairy adventures in places not yet penetrated by tourism. The railway lines were often blocked by impenetrable drifts of sand, and journeys which should have taken hours went on for days. It was necessary to put a biscuit tin to an unusual use. The worst came in a motoring journey across sixty miles of salt marsh, when again and again it was necesary to dig and pull the car out of the mire. 'At one place there was a gully 10 feet deep and about 8 feet across and we had to take off our shoes and stockings and wade through the water and mud as well as help to pull the car through. We went on till it almost got monotonous, stopping, getting out and pulling, then we came to a bit where the road was so deeply cut up that we were pitched about like shuttlecocks while the car careered about among the ruts.' At Khanga, an oasis town deep in the desert, the ladies were escorted and closely guarded by the numerous sons of the all-powerful Caid. When one of the sons bought bangles in the suk and put them on Belle's wrists she knew that she was in trouble, and told her mother nothing about that incident in an otherwise complete account of her misadventures.

While all the missionaries of the ABM were ladies, the committee which looked after its affairs in England consisted entirely of men. The name of one of these, a Mr Collinson, begins to occur in Belle Patrick's letters, and he invited her to visit him when she went on leave in the summer of 1926. It seems to have been W. Cecil Collinson (as he was always known) who suggested the new placement of Belle in Croydon. In February 1927, she had sad news for her mother. 'Mr Collinson, our Deputation Secretary, lost his wife this week and he is left with four little children . . . It is very sad for him. He is such a very nice man.' (Not all the children were 'little'. Marjorie was already thirteen, Gerald ten, Bernard seven, although Hilda, closer in age to her future half-brother, was not yet two.) In July 1927 Mr Collinson escorted Belle and another missionary lady to the Keswick Convention and proposed marriage at Friars Crag on Derwentwater. By the end of the year they were married.

I was born in August 1929, and I was not, I think, part of my father's original plan. He desperately needed a mother for his four children. This, for Belle Patrick, must have been a wholly unexpected role, although she had always loved children, from the Cellardyke bairns to the children tumbling about in the Arabic services in Algiers. When I was four and in the bath, I told my mother about wicked step-mothers in the fairy stories, and what dreadful things they did. 'But you know,' said my mother, 'that I'm Hilda's step-mother.' I thought for a moment and said: 'Well I suppose those wicked step-mothers never knew the Lord Jesus.' This precociously pious statement was often quoted.

My father was part of a large Quaker family (one of eleven children), allied to such other Quaker dynastics as the Peases and the Leans. He too had caught evangelical religion (through the Young Men's Christian Association, a rather different kind of movement from the modern YMCA), which caused a progressive detachment from the Society of Friends and involvement in several missionary causes, at home and abroad. Born in Ipswich (but of Yorkshire stock), he became a shopkeeper in Bury St Edmunds, supplying the local gentry with suits and racing stables with horse blankets. In the 1920s he took his ailing wife, Hilda Quant, on one or two Mediterranean cruises, and it was there that he felt the call to evangelise the muslim world, retiring from business and devoting himself fulltime to the home end of such missions as the Algiers Mission Band, the North Africa Mission, and a network of prayer and correspondence called the Fellowship of Faith for the Muslims.

In 1935 the Secretary of the Egypt General Mission was killed in a car crash on his way back from Keswick, and W. Cecil Collinson succeeded him. This meant a move for all of us, from Ipswich to the mission headquarters in Highbury, North London, with regular visits to Egypt for my parents. Then came the war, and the Blitz. Our house was badly damaged twice, in September 1940, and by a doodle bug in June 1944. In between, my parents ran a canteen for the forces in the Cambridgeshire fens, which determined that my own education was at Ely, followed by Cambridge. (I went to university in spite of my father's suspicion about such places, since my mother was determined that I should have the opportunity which had been denied her, and would have liked me to go to St Andrews.) After the war there was a move back to Ipswich, where my father died in November 1952.

In the twenty years of life remaining to her, Belle Collinson made house with my sister Marjorie, and continued to apply her acumen and industry in all directions. She managed the financial affairs of the family, typed (and indexed!) my own PhD thesis and everything else that I wrote until her final illness, built a new house in the garden of an old one, wrote this book, and in the 1960s went back to work in a lawyer's office in Ipswich, where she soon knew more about the affairs of colleagues forty or fifty years her junior than I would have thought possible. When she spent her winters in the Sudan, where I was a lecturer in the University of Khartoum, and where many of her old friends of the EGM were now deployed, she undertook journeys almost as exacting as those Algerian tournées of thirty years before – and she was by now in her sixties. When, in 1969, I was head-hunted for a professorial chair in Sydney, it was her decision that we should all go to Australia, my wife, my four children (the youngest not yet one), and herself. It would, she said, be the solution to our problems and to hers. It was the last of the many bold initiatives which punctuated my mother's unusual life. She celebrated her seventy-fourth birthday on board ship. In Sydney she was soon doing all the secretarial work for the local office of the Middle East General Mission (the post-Suez successor to the EGM). She died of cancer, patiently borne, in January 1972.

Belle had shared with her sister Rita the burden of the last and lingering days of her mother, who had died in July 1956 aged ninety-six. She remained close to Rita and paid her last visit not long before embarking for Australia. Auntie Rita read this book and took exception to many things in it. 'I'm afraid I'll be scoring out a good deal but I'll read and re-read it.' After all, she had stayed at home with Mother, while Belle had gone off into another world and had married a fancy English husband. What did she know about the old days in Cellardyke? That is why this book lay in a drawer for many years after Belle Patrick's death, and for some time after Rita, the last of the family, had also departed. Re-reading this remarkable book, I think I always underestimated my mother.

Patrick Collinson, January 2003

1. Introduction

THIS IS NOT AN AUTOBIOGRAPHY, though it may be that the first person singular will occur rather too frequently. Although my personal life can have little interest for anyone outside my immediate family, the first thirty years of that life were spent in a community which has now ceased to exist in the form I knew, and the life of that community was so individual, so independent, so different from the present-day standardised pattern, that it deserves to be put on record.

This is no work of scholarship. I have neither the time, the opportunity, nor the ability to make researches; all I can do is to put on paper the memories of my childhood and youth.

I was born and lived among the fisher folk, but I was not *of* them, and that, I think, was an advantage. I saw things as an onlooker, and even as a child I could note how the life of a fisher laddie or lassie differed from the life of the country folk or the townspeople. At school, in church and Sunday school, at play and, most of all, in the home, the three sections of our small society were as separate as African tribes; we necessarily mingled, but we never mixed.

We even had three separate townships – all Burghs – with not even a notice to show when one passed from Cellardyke, where the gables of the houses had a door under the roof with a pulley over it for hauling up the nets, to Anstruther, with its shops, offices and banks and a corn market every Friday, and over the Dreel Burn to West Anstruther, where the golf course and a few bathing huts on the beach at the Hind provided recreation for the artisans and an attraction for summer visitors.

And of course each of the Burghs had its own school, so that it was not till we reached the 'Qualifying Standard' and – for most of us only if we had brains enough to win a bursary – got into the Waid Academy, that we all got together: the fisher bairns from Cellardyke and other neighbouring fishing villages, the families of shopkeepers and tradesmen from Anstruther, and the farmers' sons and daughters from the country. We were all together in class. In the main the fisher bairns were the brightest; they were all winners of bursaries, whereas the farmers' children were almost all fee-paying, and the others were betwixt and between: but in the playground and sportsfield we inevitably went to our own company. I had the misfortune, or perhaps it was a privilege, to belong to none of the three groups. My father, who had died about three years before I won my bursary to the Waid, was an employee of one of the firms of fish-buyers and therefore worked with fishermen, while my mother's father was the harbour clerk and she was brought up among the fisher folk.

In religion too the division was marked, and in this sphere also I was among the 'dykers'. My father and mother were members of the Baptist Church – the 'Dippies' in local lingo, who were mostly fishermen and their families. But about the time I was born the head of my father's firm, who had for some time been precentor in the Parish Church (or Auld Kirk) in Cellardyke, wished to retire, and, knowing that my father possessed a very fine tenor voice and a true 'ear', asked him to take over the job, which carried an honorarium of £5 per annum. This was the year's rent of the house my parents occupied at that time, and with seven, or it may have eight by that time, children my father did not feel free to refuse the offer, though he knew that it would mean excommunication from the local Baptist Church and alienation from many of his friends. He told my mother that she was free to go with him or to remain, but mother decided that a division in the family was more to be feared than a change of denomination, so she and all the family, except my eldest sister who has remained a loyal Baptist all her life, walked past the Baptist Church to Cellardyke Parish Church. At that time denominational feeling ran high, and to the Baptists no other denomination had a clear title to the name of Christian, and the Church of Scotland least of all. So, as the little family filed past the open door the Baptist deacons, so recently their closest friends, remarked to each other in voices loud enough to reach across the street: 'There goes Alec Patrick leading his bairns to hell!' The superiority of the Baptists to all others (in their own eyes at least), dies hard. Not so many years ago, a friend of our family, a godly old woman who has served her Lord in the Church of Scotland for over fifty years, joined a party of Baptists who were attending a rally in a nearby township. She was introduced to some of the ladies of the church where the rally was being held as 'Mrs —, not a Baptist, but a Christian, which is the next best thing.' And this in all seriousness!

When I left school I started to serve my time as shorthand-typist in a local solicitors' office – four years at a progressive salary of £5, £7 10s, £10, and £12 10s per annum. The chief clerk was secretary of a local mutual insurance society run by fishermen which insured all the fishing boats, and as his assistant I gained valuable insight into the principles on which the finances of the fishing industry were based. It is illuminating that the company went into voluntary liquidation, not because of financial stringency, but because no one would sit as a director and thus become the target for bitter personal attacks on the part of any skipper who found himself aggrieved over the settlement of a claim for damages. At the end of my four years' apprenticeship I was promoted to being in sole charge of the Town Clerk's Office of Pittenweem, a place which was then at the centre of a predominantly fishing community. I began my duties there on 1 August 1915, and the registration date for the first national register was 15 August 1915, so within the first month I had details of all the 3,000-odd inhabitants of the Burgh. This was followed by the first of all rationing schemes, and again I had to issue ration cards to every family. Military Service Acts followed in quick succession. Most of the men in Pittenweem were in reserved occupations – fishermen or farmers and farm labourers. When the single men were called up, few of them appealed against military service on the ground of their occupation; a great many were already in the forces, for in 1914 almost all the drifter fleet had been commandeered to serve as minesweepers, and with them went most of the young fishermen. Young farmers joined the Yeomanry in 1914 and most of the young labourers and working-class men had already joined the 'Territorials' or the Black Watch and were in France by the time of the Battle of Mons. But it was a different story in 1916 – May or June I believe – when the older age-groups began to be called up. Most of the local fishermen were up North, fishing from Peterhead, and it was their wives who came to the office to ask for application forms for exemption and for help to fill them in. It was then that I got to know these women and began to realise that the fishing community was a matriarchal society. It was the women who did all the business in the family. I remember one fisherman, a man in his early forties, who had a family of five or six children, and who was fishing at home when his age-group was due to register. He came up himself for the form and started to give me the information I required to fill it up. He knew his own full name and that of his wife, but he could not remember the date of his marriage nor the dates of birth of any of his children. 'The wife attends to a' that', he said, and having signed his name to an almost blank form, he went home to send his wife up to the office to attend to the rest of the business.

Coal rationing was the next imposition, and once again it was the fishing community on which it bore hardest. Not only was domestic coal rationed, but steam coal, without which the few old steam drifters, not fit for naval service, which had been left to carry on the fishing, could not put out to sea. This time I had to deal with men. Whereas the women had been grateful for my help and became friends, the men, probably resenting having to be dependent on a young girl for vital supplies for their boats, the only things in their lives which women did not run nor meddle with, were angry and resentful, and, as one angry skipper told me, they held me personally responsible for keeping a drifter in the harbour, while others, who had been fortunate enough to get supplies in one of the larger ports, were making big money.

Then came 1918 and the Representation of the People Act, with 15 June (so far as I remember) the qualifying date for the vote. No voters' roll had been prepared since 1914 and it was one of the duties of the town clerk (and therefore mine) to prepare the first roll under the new Act. There was no one in the office who had ever prepared such a roll or had anything to do with its preparation. I was handed a copy of the Act – the first Act of Parliament I had ever seen printed complete and separately – and told to prepare the roll. So I began with the preamble, read through every word of sixty-three long clauses, making voluminous notes on the qualification of voters and all else covered by these clauses, and painfully came to Clause 64, which began 'The foregoing provisions of this Act shall not apply to Scotland but in lieu thereof the following provisions shall take effect', or some such words, and then followed as sub-sections to Section 64 the whole of the Act as it applied to Scotland. That was the first and only time in my life I had any sympathy with Scottish Nationalism.

I can still remember the sheer drudgery of compiling that first register. I had to do it in the evenings, since office hours were already more than full with the routine jobs of town clerk, burgh chamberlain, collector of rates, and all the administration of the wartime National Register, food and coal rationing, local tribunals, and so on. In the case of Pittenweem the comparatively simple arrangement of some 3,000 names in alphabetical order was complicated by the fact that, as in most close-knit communities, the same surname was common to so many families. Andersons, Wilsons and Watsons were numerous, but the letter H, as far as I can remember, took up one quarter of the whole register, since there were so many by the name of Horsburgh and still more called Hughes. There was also a purely local custom whereby a married man took as an additional surname, written in brackets after his own surname, the maiden surname of his wife. This was the family name borne by all the children until they married, when the bracketed surname was altered, in the case of sons, while in the case of daughters the maiden surname became the bracketed surname of the new family. To find one's way through families of Hughes (Hughes) who became Hughes (Horsburgh) or Horsburgh (Hughes) became a nightmare, but by the time I had finished I knew most of the ramifications of the Burgh, where intermarriage was the rule and the exception of marriage with an 'incomer' practically unknown. Each family, too, had its particular Christian names which were repeated in each generation. These were for the most part the common Christian names: Alexander, Andrew, David, James, John, Thomas and William for the boys, while the girls were Agnes, Helen, Isabella, Jean, Margaret, Mary with all the masculine names adapted by the affix of 'Ina', which was the only part of the name used; but the clan of Horsburgh was further distinguished by the fact that in every family one boy was named Lock and one daughter Preston.

This register, so painfully prepared, was subjected to the test of having a general election fought on it. Pittenweem was one of the group of 'St Andrews Burghs' which was merged into the constituency of East Fife. East Fife was Mr Asquith's seat – absolutely safe, solidly agricultural and held (so the Tories said) by the Liberals under Asquith by the simple plan of promising at each election to give the farm labourers a half-day on Saturday and forgetting all about the promise from one election to another. St Andrews Burghs, on the other hand, believed in close-fought campaigns which resulted in a sort of seesaw in which neither party achieved more than a three-figure majority, and sometimes not even that. It is a matter of history that Asquith was defeated and had to be found a seat elsewhere, but as far as I was concerned it meant that practically every person on the register recorded his or her vote, and omissions which would have gone unnoticed perhaps for years were brought to the notice of the town clerk responsible, with bitter charges of inefficiency. In the circumstances it was not surprising that there were glaring mistakes. In one district in West Fife it was reported that a whole street was omitted. I was therefore hardly to be blamed for omitting a certain Wm Wilson in favour of his son, Wm

Wilson (a) ('(a)' stood for 'absent voter') who was still serving when the register was compiled, but blamed I was and in no uncertain terms by Wm. Wilson Senior, who made a special call at the office to tell me what he thought of my political manoeuvring. He was a Liberal; I was known to be a Tory, and who knew of how many more votes I had deprived Mr Asquith!

Then came the first Housing Act, preceded by inquisitions into the housing conditions of the Burgh. Reams of forms had to be filled up with statistics of accommodation of each house, number, sex and age of the inhabitants thereof, and so forth. When the forms had been submitted to and assimilated by the appropriate government department in Edinburgh an inspector was sent down to check up. Fortunately for himself he called at the Town Clerk's Office first, and discovered that I could give him all the information he needed about every household in the burgh. On the doorstep as he was leaving he turned to thank me, adding: 'I think there is only one thing you have omitted. I still don't know what style of corsets the women wear!' He afterwards sent me a little suede-bound volume of Stevenson inscribed with thanks for my invaluable assistance.

Well, of course, I have said far too much about myself, but in this introductory chapter I thought it was necessary to show what I consider to be my qualifications for writing this book.

2. Religion and Morality

RELIGION AT THE END OF THE nineteenth and early years of the twentieth centuries a really vital part of community life, and in this department perhaps more than in any other the difference between the fisher folk and the country folk was most marked. The tradespeople occupied a place between the two. The country folk (grieves, ploughmen, carters etc.) were usually attached to a country church where the religious exercises of the week consisted of one service on Sunday, followed by a Sunday school. In the fisher villages, on the other hand, church life was more varied and much more vigorous. Even the Auld Kirk in Cellardyke was in fact a new church, being a *quoad sacra* parish created through the efforts of a local bank agent who realised that the staid, solemn afternoon service in the parish church of Kilrenny could not hope to keep within the Church of Scotland the strong, virile Christian men and women of the fisher community, in competition with one congregation of the Free Church of Scotland, one of the United Presbyterians, one belonging to the Evangelical Union, and a Baptist church, all of which had a full programme of two Sunday services, Sunday School and bible class, all on Sunday, with prayer meetings, guild meetings, Band of Hope and temperance meetings on week nights. So Cellardyke Parish Church came into being, and provided all these things, and it was in that church, whose congregation was largely composed of fisher folk, that I was brought up.

So much has been written about the dreary Scottish Sabbath and the disastrous effect on the young people of being forced to attend services and Sunday School that I must say that in my own experience I was completely unaware of any such thing. I and most of my generation regularly attended church twice each Sunday and many of us also attended two Sunday Schools, but in our home at least Sunday was a very special day, the one day in the week when mother had leisure to play with us. Looking back, I marvel that this was so. Today not many mothers of ten children could manage to turn the whole family out to church at 11 a.m., leaving a dinner cooking so that we could all have a hot meal and be back at church for afternoon service at 2.30 p.m., home for tea, and the children off for five o'clock Sunday School. Not much leisure in that programme. But leisure is not measured by hours or minutes; it is a state of mind, and Sunday was a day of rest, when there was no rush. Nobody said: 'I've no time to be bothered with that now', and after Sunday School, when father and the 'big ones' had gone off to a third service somewhere, we 'young ones' had mother all to ourselves, a mother with no knitting or mending in her lap, no cooking to be put in hand against the morrow, no ironing – nothing to occupy her but five small children. When I look back I realise that there was not much physical rest for mother even on the day of rest, but the happiness and peace of all those Sabbaths live on in my memory, and I am sure in the memory of many more who were blessed with such an upbringing.

In addition to this full and active church life, there was among the fisher folk a strong tradition of religious revival. In each generation it seemed there had been such 'times of refreshing' when, sometimes spontaneously and sometimes as the result of the visit of some travelling evangelist, the Spirit of God manifested Himself in a way not known at other times. My grandfather used to tell of the revival in his day, when strong fishermen, not notable for 'good living', as the local phrase was, were struck down on the way to the boats by conviction of sin and were physically unable to move from the cobblestones of the streets or piers until someone could kneel with them and bring them assurance of salvation: and a less emotional man than my grandfather you would have to go far to meet.

It was in a revival of this kind, brought about by a visit from two young women evangelists of the Faith Mission that my father, an alcoholic in the making, was soundly converted and became himself an evangelist in his spare time, taking a special interest in the nomadic 'tinkers' who had regular camping places on the Marches between the parishes. In these wide grassy lanes, between high hawthorn hedges, a no-man's-land between jurisdictions, there was a continual but ever-changing population of tinkers, with their tilted carts and primitive tents, and there my father, with his clear tenor voice, sang and preached to his wandering congregation.

The revival I remember most clearly in my childhood must have been in 1906 or 1907. The evangelist was a fisherman, or at least of fisher origin, from north-east Scotland. The meetings were held in Cellardyke Town Hall, so that they were free from denominational label, and from the start they were successful. The time was well chosen. At the end of November or beginning of December the boats came home from the 'South-fishing' at Yarmouth or Lowestoft. The winter herring fishing did not start till January, and though the days were busy enough for the fishermen in overhauling boats and gear, the evenings were free. So night after night the town hall would be filled with two or three hundred young fishermen, and where the lads went the lassies were sure to go. The younger clan of school-children filled the front benches with an excited clamour, heads mostly turned round so as to be able to recount next day at school how many 'held up their hands' and whether they wept or not. Fisher folk are great singers and the revival hymns with their unashamed, sentimental, emotional words and catchy tunes did their full part in preparing the audience for the address. There would be a few prayers offered by some members of the congregation, personal heartfelt petitions for blessing, not on mankind in general, but on that particular gathering, and specific in their application. The bible-reading was traditional but perfunctory, something to be got through before the real business of the meeting began. As I remember, the evangelist had little 'education', and though he usually began in stilted 'English' (certainly not recognisable as such by anyone from South of the Border) he soon lapsed into his mother tongue, to the great satisfaction of all his hearers. I have no recollection of the speaker's style; indeed I did not know there was such a thing as style in preaching, but I can still recall the entrancing racy way in which he carried us along with him. There had to be a text, of course, but it was invariably out of its context and was only used as a peg on which to hang the outline of the address. I remember one on 'And David's place was empty' which had nothing to do with David's escape from the evil designs of Saul, but set out in vivid word-pictures the tragedy of an empty place in the home, in church, in the lifeboat on a mission of mercy, at the Great Wedding Feast of the Lamb. Here was no expounding of scripture, no 'thus saith the Lord' or 'the Bible says'. It was an open and calculated appeal to the emotions, and as such it succeeded. After the address the speaker asked those who wished to make a decision to indicate this by raising a hand 'while every eye is closed and every head bowed'. There may have been bowed heads in the front rows, but they were not turned to the platform, and behind the shading hands were wide-open eyes carefully noting every hand raised, every tear or other sign of emotion, all to be recounted to admiring groups in the playground next day. After the meeting proper was dismissed, there was an 'after meeting' which all those who had raised their hands were expected to stay and attend, in order to be counselled by the evangelist or by experienced Christians who remained behind for that purpose.

Along with my brothers and sisters I attended these

meetings regularly, though we did not sit in front with the other children. As the days went by more and more people, including children, were 'saved', and soon the playground at school held two separate camps – the sheep and the goats. I hovered unhappily between the two groups: of course I was not a goat; my father had been an outstanding figure, and all our family were 'good living' so I could not be 'unsaved'. But I could not recount how long I had sobbed, how many tears I had shed in the reading room at the 'after meeting', so equally I could not claim to be 'saved'. One evening I went to the meeting with my heart set on one thing, and one thing only. I would be 'saved' and join the elect. I have no recollection of the meeting at all, but I remember every detail of the 'after meeting'. I was 'dealt with' by the evangelist himself and so was greatly privileged and would be able to boast of this at school next day. But I would have to admit that, try as I would, I could not produce a single tear, and that I knew would bring grave doubt as to the reality of my conversion. But this I do know, that never since that day, now nearly sixty years ago, have I had any doubt that I was then born again. What I did then was definitely to acknowledge myself a sinner and to accept from God the salvation he offered freely, forgiveness for my sins and His power and presence in my life to keep me in His way. Of course I was swayed by emotion; of course I took the step chiefly to be accepted by my schoolmates as one of the élite; of course not one of my motives was a genuine desire for a new life; but the result was the same. I was definitely committed to serve God and in His service I have found pure delight.

Head knowledge was one thing. We could quote the Bible for hours on end, and as for the Psalms, they were part and parcel of our everyday language. I remember being teased by the Latin master at school, who was believed by us to be an atheist and who took a special delight in mocking anyone in his class who made any profession of religion. One Monday morning he made me stand up in class and said: 'Now, Miss Patrick, will you please tell us what you taught your children in Sunday School yesterday?' I was then thirteen, and I was indeed a Sunday School teacher, though I am sure that I learned more from my class than they learned from me. However that made no difference; I always gave back as good as I got. So I replied in a meek and demure voice, quoting from one of the metrical psalms:

> Than all my teachers now I have
> more understanding far;
> Because my meditations
> thy testimonies are. – *Psalm* 119, v.99.

I was never again teased in the same way. But head knowledge never saved anybody, and what we needed and got in our revival meeting was a challenge, a call to 'step over the line', to respond to Joshua's question: 'Who is on the Lord's side?'

Morality in East Fife, as elsewhere, meant first and foremost sexual morality, but though 'impurity' was undoubtedly the greatest sin, intemperance was a close second. Among the fisher folk there was and always had been a great deal of premarital sex, and provided marriage followed, no one thought any the worse of the couple. This was in marked contrast to the tradespeople and shopkeepers. Among them, for a daughter to be 'in the family way' before marriage was a disgrace, though more social than moral. Among the fisher folk the child born out of wedlock was more readily accepted into the family too, and the fiction of 'auntie' instead of 'mother' deceived no one, not even the other bairns at school. A new teacher, unacquainted with the family history of her pupils, once asked a child what was his father's name. He made no reply, but the boy in the next seat vouchsafed the information: 'He hasna a father or mother – his auntie had him to the lodger.'

Church discipline was strictly observed, and if a child was born to a couple within the prescribed period after marriage, they both had to 'go before the Session', before they could take Communion, but in practice they either did not join the church before marriage (the usual time for young people to 'come forward') or they did not come to communion until probably several children were in the family and their ages were not likely to be closely scanned. When either 'guilty party' had scruples and wanted to put matters right the session meeting was an embarrassment to all concerned, as the elders were in all likelihood close friends or relatives of the parties. Each was expected to say a word of rebuke and counsel to the offenders, and it is reported that one elder, faced with this task very shortly after his ordination, could only stammer out: 'Well, well, it's past and done with. Just see that ye never do it again.'

As to grosser immorality, no doubt there was as much among the fisher folk as elsewhere, but such things were not openly discussed when I was young, and so I have no facts to go by. Once I remember my mother saying that my father had been told by English fish-buyers that there was more sin and wickedness in our town than in any other they visited. But the Englishmen knew that Father was an evangelist and it may be they wanted to taunt him both with his faith and nationality.

Drunkenness, of course, could not be hidden, since public houses were open from 8 a.m. till midnight, and fishermen who 'liked a dram' would often be drunk in the streets before dinner-time. They would come in to the harbour in the early hours of the morning, cold and hungry after a long night fishing, and all round the harbour were the welcoming doors of the pub. So many went in for a dram and some would stay till dinner-time when we children would see them staggering home to get sobered up in time to go out to the fishing again on the evening tide – a hard life and hard men who lived it.

If sexual immorality was the unmentionable sin, drunkenness was the subject of open warfare. In the fishing community the battle line was clearly drawn. On the side of the angels were the various temperance societies. The most spectacular one I remember was the IOGT; the Independent Order of Gospel Templars was, I believe, the full title. As children we interpreted the initials as 'I Owe God Tuppence', and we felt terribly wicked and daring when we whispered the blasphemous words to each other. The great campaign against the 'demon drink' had passed its zenith before the twentieth century began, so I have only hearsay to go on as far as the IOGT is concerned, but I believe it was a society modelled on Freemasonary with rules and ritual only revealed to initiates. I can remember my brothers mimicking an old man who was Chief Templar or something of the kind, as he opened the session with the ritual sentence: 'We are here to work; let us do so, and so fulfil our oath and honour God.' Never was an unwelcome task undertaken by one of us but our older brothers 'encouraged' us by chanting these words.

The society I knew most about was the one in which my father was one of the leaders – the Gospel Temperance Union. This society too had its signs and tokens by which members could recognise each other – a piece of blue ribbon on the lapel of a man's jacket – but there was nothing secret about its meetings or its aims. Temperance by itself was nothing. A man might be delivered from drunkenness and be little the better for it, and they believed that such outward reformation would never last. Only the power of the Gospel could overcome the 'demon drink' and deliver his victims to a life of victory and active aggression against their old master. And to achieve this end, every means was to be used. To combat the attraction of the public house on a Saturday night, there was 'The Saturday Nights Meeting'.

I myself was too young to attend one of these meetings; when I was nine I lost my father and we moved away from the borders of Cellardyke to the snob-land of Wast Enster, and so for a time I lost close contact with the fisher folk. By the time I got in touch with the 'Saturday Nights Meeting' again it had lost its zeal and evangelistic fervour, and especially its out-and-out aggressive attitude to the drink question and had become just one more weekly gathering of the 'good living' where people congregated to sing rousing hymns, to listen to solos and recitations by local talent and to hear a mediocre address by one or another of the committee members. This was very different from the hearsay reports I had heard in my childhood when real drunks would come in, or be fetched from the nearby pubs, be sobered up with cups of strong, sweet tea and rewarded with a juicy meat pie for listening to impassioned appeals by ex-drunkards, who could testify to the benefits of salvation.

Of course the children were catered for too, and every Wednesday evening we all made our way to the Band of Hope at the Baptist schoolroom. An aunt of mine had the misfortune to live in a flat above the schoolroom and her comment on the meetings, repeated whenever they were mentioned in her hearing, was: 'Band of Hope – band of fiends, more like!' It is difficult for youngsters nowadays to imagine the world we lived in – no radio, no television, no cinemas, no organised entertainment of any sort. In those far-off days we relished all the adventure of being Christian Commandos (though that term had not then been invented) as we shouted: 'Dare to be a Daniel, dare to stand alone', and the hundred or so of us would gladly have faced lions or drunk men when we knew so confidently that God would shut their mouths. It was rather different when we separated outside the hall and went off alone to our homes. Mine was up a narrow 'wynd' with gaping cellar doors to be passed, each one of which could, and quite often did, shelter a drunk who had been turned out of the pub at the corner. My method of 'daring to be Daniel' was to shut my eyes tightly, repeat the Lord's prayer under my breath, and run in the middle of the roadway as fast as my legs would carry me. It was just as well that there were no cars or motor cycles in those days. The only road drill we learned was concerning the broad and narrow ways and the Rule of the Road was: 'Keep in the middle of the King's Highway'! So I always reached home in safety. Many years later I was going up that same wynd and I found myself saying the same old prayer.

It was one thing to attend the Band of Hope; it was quite another to 'join' the Band of Hope. Any child could go to the meetings, could enjoy shouting the choruses, could see the magic lantern shows, where the story inevitably began in a slum home with a drunken father knocking about his long-suffering wife and terrified children, and always finished with the same slum room transformed within a week of the father's conversion to a spotlessly clean and beautifully furnished home with mother and children in new clothes sitting down to a substantial meal while the father said grace.

Even in my young mind such pictures raised questions as to the economics of such a transformation. But one could only 'join' by 'signing the pledge'. This was done at a special meeting, when any child old enough to read and write might file up to the platform, sign the fateful form, receive a certificate, and become entitled to wear a scrap of blue ribbon to show to all the world on whose side he or she was in the 'Great Fight'. My turn came and I presented myself before the leader, who placed his finger on a line and said: 'Just put your name here.' I believe all the others did just that, but since the day I had first made sense out of printed symbols I could never resist reading whatever came under my eyes. So, even as he spoke, I had been reading the things I was about to pledge myself never to do. Somewhere about the middle some words appeared to stand out in red: 'Nor ever to carry intoxicating liquors'. Now I would gladly have promised never to taste intoxicating liquors under any name or description but 'to carry' them was something else altogether. My grandmother was an old lady who, like many old ladies of her day, was an invalid confined to bed by a 'bad heart'. This 'bad heart' had to be stimulated on occasion by a teaspoonful of whisky so that, in this household where the battle against the 'demon drink' had been waged incessantly for half a century or more, it was now necessary to keep a small bottle of the deadly stuff. I did the shopping for my grandmother's household, and I knew only too well what was in the bottle which was filled and wrapped in the back shop of the licensed grocer and discreetly slipped into my basket among the innocent groceries. No, I could not promise 'not to carry intoxicating liquors', so I shook my head and stumbled off the platform, shamed before all my friends but saved from the sin of making a promise which I had no intention of keeping. Yet I daresay I have kept the pledge more faithfully than most of those who signed on that occasion.

I don't think there was much, if any, social drinking among the fisher folk in their homes; at least I never came across any, except, of course, at Hogmanay, when everyone, except the 'teetotallers', served drinks to anyone who came 'first-footing'. Drinking was done in the pubs, and only men frequented pubs. There were several old women who used to hang around the public houses at opening time (8 a.m.) rubbing a threepenny piece (the price of a nip of whisky) between their fingers. If anyone they knew happened to pass, the reply to a greeting was always the same: 'My clock stopped in the night and I've just come down to look at the Town Clock.' No one was deceived, but the decencies were preserved. Women did *not* frequent pubs; these were lone women, widows, who had no man to give them a dram when they needed it, so they had to go to the pub.

To me as a child the public house was the visible 'Seat of Satan'. I had to pass many on my way to school each day, and I always did so keeping well to the edge of the pavement or even walking in the gutter lest any evil should befall me. I remember especially one very respectable 'Inn' which was run by three sisters, middle-aged maiden ladies. One day, on my way to school, just as I passed the door, one of these ladies looked out, and seeing me asked if I would go on an errand for her. I had to say 'Yes' for no one ever refused such a request; all passing children were accustomed to do such little jobs for anyone who asked. But instead of handing me a note and the money she asked me inside and showed me into the bar parlour, while she went to the back premises to get what she wanted to give me. I literally shook with terror. The wide window on to the street was shaded by a wire blind bearing the words 'Bar Parlour' and I could see through it quite clearly. I did not know that from the street side it was opaque, so I shrank into the farthest corner of the room and stood with my face to the wall, hoping that no one would discover my disgrace.

This aversion to being associated in any way with a public house was not mine alone, nor among the things that changed suddenly after 1914. When I came to live in Ipswich in 1928 a young girl of my acquaintance, whose father and mother were both 'good-living' fisher folk, accompanied her employers, a retired schoolmaster and his wife, from Anstruther to Ipswich to settle them in their new home. She naturally wanted to look up one whom she knew very well as a small child and asked for directions to pay me a visit. She was told to take a tram to a certain corner, to alight there and ask to be directed to the Woolpack Inn and there to take the right-hand fork which would bring her to my home. After hours in which she must have explored a great deal of Ipswich, she turned up, almost in tears. I asked why she had found it so difficult to find the house, and she told me: 'Mr McIvor told me to ask for the Woolpack Inn, but I would be black affrontit [deeply ashamed] to ask for a public house, so I just kept on looking.'

Gambling was another pursuit which was anathema in my home, but was widely carried on among the fisher folk, both young and old. I never knew if it was legal or not, but as we played round the piers and beaches, we would come on groups of boys or men playing card games with piles of coppers which I presume were the stakes. Playing-cards were 'The De'il's books' and we never lingered near these 'De'il's schools', so I don't really know much about these activities. So far as I know this pastime also was exclusively a male one, but the women may have had their games of chance, though I hardly think many of them would have found leisure for such a thing.

I don't remember ever hearing of betting on horses while I was a child. That of course means nothing at all, but it is strange to realise, on looking back, that my first consciousness of this activity was brought about by such an innocent-seeming luxury as ice-cream. The first ice-cream I remember was sold from brightly painted hand-drawn carts by Italians who were called 'Hoky Poky men'. These were apparently harmless people, because, though we seldom had the necessary penny to buy an ice-cream 'slider' [wafer], we were not forbidden to do so. But it was a very different story when one of these immigrants rented business premises, brought over his wife and children, and started an ice-cream shop, as brightly and outlandishly coloured as the old hand-carts. Immediately an absolute ban was pronounced on what our elders called 'the De'il's infant school', and in whispers in the playground we discovered that the back shop was used as a betting-shop. What truth was in that rumour I know not, but it is certain that the ice-cream shop became the place where the young men and boys gathered, where the girls who wanted to meet boys contrived to pass every few minutes in the promenade of the street and to which the 'good-living' girls gave a wide berth.

The theatre was another abode of wickedness, though in our little town it was a very temporary one. Only very occasionally a travelling theatrical company would make a one-night-stand in the town hall. What size or kind of audience they drew I had no way of finding out; we were not allowed to go near the town hall even during the day, though the performances were always at night. But on two occasions we had a fleeting contact with the theatre. One company had need of a chorus of children, and the manager came to the school to recruit his choir. We as a family were known as singers, and so naturally we were put forward, and one of my sisters actually 'signed on' on the promise of a fee of one shilling – a fabulous sum to us. The rest of us said we would ask Mother, though we knew quite well what her verdict would be. Poor Annie had to go and withdraw her services and so she lost any chance of riches. I can remember Mother's real horror at the thought of one of her children being associated in any way with the theatre.

It was a younger sister who touched the fringe of the theatre in the other incident. *Uncle Tom's Cabin* was being staged for three nights and there were real negroes in the cast. That itself was of great significance, as it was the first time most of us had ever seen anyone of a different colour. But one of the negro women, who was lodged in a house near us, was brought to bed of a child. Rita, my sister, had never been able to resist the chance of nursing a baby. She was fair game for any mother who wanted an unpaid nurse for a fractious baby, and she seemed to have an extra sense which guided her to any infant. So the midwife had hardly finished her work when Rita, aged about five at the time, appeared in the room and took the infant in her arms. To have so young a baby, and a black baby at that, was absolute bliss, and no one seemed to worry in case she should drop her charge. So she had time to make a minute examination of the tiny piece of humanity and was able to pass on to us the fact that the palms of the hands and soles of the feet were pink and not black at all.

Swearing or the use of any word which in those puritanical days had to be printed as —— if it was printed at all was another sin with which we were all familiar. 'Fishwives' language' has always been blunt, to say the least of it, and the men were even more outspoken. We all picked up the words, of course, and I have no doubt my brothers and sisters used them outside the house in common with all the children of the neighbourhood. But for me, personally, it was a thing I could never bring myself to do, even when I wanted to be among the ringleaders of our gangs. I don't know why this was so, but I do remember as a very small child being terribly puzzled when I was told that it was very wicked and a sin to say certain words. The oaths 'By God', 'For Christ's sake' etc., I could understand; there was the commandment: 'Thou shalt not take the name of the Lord thy God in vain for the Lord will not hold him guiltless that taketh his name in vain.' That prohibition was clear and absolute. But why should the other words, the meaning of which I had not the slightest clue, be wicked and sinful? They were just words, and how could words in themselves be wicked? One day, when some small fisher boys were using every forbidden word in a gleeful parade of bravado, I crept away to a secret place in the boatbuilders' yard and tried to think it out. In real agony of soul, though I was only about five years old at the time, I tried to find a solution, and suddenly I remembered the scene of the crucifixion and the men who 'railed' on Him who was the central figure. I liked my stories to be precise, and I had often wondered what one would say in 'railing'; of course I had the answer to my problem: it was these forbidden words that the men used in 'railing' my Lord. Never since that day have I used such a word even in thought, and the result is that my vocabulary contains none. I really have no idea what 'four-letter words' are; by closing my mind against them I never even heard them from that time, so I can say nothing about this 'sin'.

Truthfulness and honesty were so much a matter of course among the fisher folk that they were seldom questioned. There were of course the liars and rogues but, so far as I remember, they were well-known and their failings were regarded with the same pitying tolerance as the 'dafties' or 'half-wits'. A newcomer to the district might be taken in by a plausible tale, but 'incomers' were fair game in any case, and if any complaint were made the reply would be: 'Oh, ye canna believe a word he says, puir man.'

I should think a certain amount of poaching went on, both of salmon and game, because these were never named openly: a salmon was always 'a red fish' and pheasants or partridges were always called 'broon do'es' [brown pigeons or doves]. But I do not remember ever hearing of a prosecution for poaching.

3. Home Life

THE HOMES of the fisher folk were completely different from the other houses, both architecturally and in their furnishings. They consisted, like all poorer middle-class dwellings of the period, of a 'but' and a 'ben' – two rooms only, but whereas the 'cottar houses' of the ploughmen were single-storied cottages with a room on either side of the door, the fishermen's houses were usually tenement dwellings built in terraces. The street door opened into a long dark concreted passageway, and the two rooms of the ground floor dwelling opened separately on to this; beyond both doors the staircase went up steeply, built in and also completely dark. At the far end of the 'trans', as this passageway was called, was the door to the back premises and thence to the garden. Only by leaving one or both doors open was there any light in the 'trans', and as there was usually a strong wind blowing, such an indulgence on the part of visitors was not encouraged. I got to know these houses very well as a child, for one of my duties was to take round *Bright Words*, to the subscribers to this monthly evangelical magazine.

After one negotiated the stairs, the first-floor house was exactly the same as the ground floor, except that a tiny scullery was contrived over the 'trans' and opened out of the kitchen. Downstairs there was a sink in the window. Again there were dark stairs, even steeper than the others, which led to the garret. This was a low room which occupied the whole space under the rafters. It was well floored but the rafters were seldom covered and at each end there was either a window or a 'bole' [hatch] with a pulley used to haul up the nets and other gear. For the garret was the workshop, shared by the occupiers of the other floors. It was here that the women mended the nets and here the nets were stored between fishings. It was the place where courting couples took refuge when the weather made outdoor trysts impossible, and many of the children who were born within the first six months of marriage were conceived on the soft piles of nets in the warm scented (or at least odorous) darkness of the garret.

At the far end of the 'trans' was a paved or concreted yard out of which opened the usual domestic offices of the period: the closet, 'dry' or 'water' as these amenities developed, the wash-house with a large 'boiler' [copper] and a shed where vegetables and herrings were stored and fishing gear sorted out. Salt herrings were a staple of the diet. In the corner of the yard was another boiler which was used for boiling the nets. The winter nets were brown and boiled with 'kutch' or bark, and the summer nets were white and boiled with alum. The brown nets had a warm, aromatic odour, but the reek of boiling alum was, to me at least, a most nauseating smell from which it was impossible to escape during the summer months, for there were communal boilers in yards round the harbour which were kept busy for the whole of the summer herring season, which coincided with the school holidays.

To return to the homes, these were almost identical in furnishing in all the fisher folk's houses. In each of the two rooms were two 'box' beds. These were constructed from a base of boards, about three feet from the floor, built into the recesses in the wall opposite the window. On the base was laid a mattress filled with chaff, and over that again a feather-bed. The recesses were curtained off with whatever was fashionable for curtains at the time, and a valance hung from the bed to the floor. In the kitchen there was a kitchen range for cooking and heating, and an eight-day clock of the 'grandfather' type stood in front of the partition which separated the two beds. A white wood table covered with a chenille cloth stood in the middle of the room, and on either side of the fireplace were armchairs of varnished wood, with perhaps a thin cushion on the seat, while ranged round the walls there were six chairs of the same pattern without arms. On one wall was the kitchen dresser with its fine array of china, and on the mantelpiece and on every other available shelf were ranged 'ornaments'.

Fashion in such ornaments was even more important than fashion in dress, and the phrase 'Dyker ane, dyker a', which I heard so often in my childhood summed it up. If 'tingaleeries' were the fashion, not a respectable fisher house would lack a pair of these tall, brightly coloured glass vases with out-curved rims, from which hung six or eight cut crystals which caught the light and dazzled the eyes when any heavy footstep jangled them – hence the name. No jewels of the orient were ever coveted more than those ornaments which I glimpsed each time I handed in my *Bright Words* and received the penny in exchange. But alas, we were not fisher folk, and such riches were not for the likes of us. There were also china dogs in pairs, but my grandmother had a pair of those, so they were more familiar and therefore less desirable.

The local doctor, who was a frequent visitor in these homes, used to rave at the dust-collecting rubbish and advise the women to take them to the beach and let the laddies make a 'cock-shy' of them. He lived to see just that take place, not for sanitary reasons (for indeed such a disgrace as a dusty mantelpiece would not be tolerated in most of the homes), but simply because fashion changed in favour of bronze horses (plaster figures covered with bronze paint), and to make room for these ramping monsters, each held in check by a bronze athlete, the old 'tingaleeries' had to go. Had they been stored in a corner of the garret instead of being smashed on the beach they would today have been valuable pieces of Victoriana.

The 'ben' end, as the other room was called, had, in addition to the two box beds, a suite of two armchairs and four upright chairs, upholstered in haircloth, which made the state occasions on which one was allowed into this sanctuary dangerous and painful to small legs which did not reach the floor but seemed to find out every place where the hairs had ceased to be cloth and became instruments of torture. If the family were musical, and most fisher families were, there would be a harmonium or American organ in a very ornate case, with a mirror set in the back to match the mirror in the overmantel over the fireplace, and on each there would be the inevitable 'ornaments' in the latest style.

It may be of interest to relate the provenance of these ornaments, and where the fashion decrees originated. It may be some satisfaction to the dour Scots who have no room for such gaudy trifles to know that they came from England – the South to the fishers. The 'Sooth fishing' was from September to the end of November, when the boats fished from Yarmouth and Lowestoft, and at the end of the fishing the boats sailed north, laden with every man's 'Sooth' for his wife and bairns. No need for Santa Claus for these fortunate children; Faither brought apples, oranges, grapes and nuts, rock and other sweeties. For the womenfolk of whatever age the traditional present was china – tea sets for the girls' bottom drawers, 'ornaments' for the wife, so that she could be 'upsides' with other wives, and smaller 'mindings' for the grannies and old-maid aunties. So it was Gorleston and Lowestoft shops or stalls that decreed what should make the homes beautiful in the far-off Scottish fishing villages. I remember too, at the end of the 'summer herrin', which was a home-port fishing, travelling salesmen came to the town and held auctions of cheap china, but as the end of that fishing coincided with the harvest, when the country folk also had money to spend, I don't know whether the fishers or the ploughmen were the bidders at the noisy auctions held under glaring naphtha flares.

The floor coverings were almost invariably shiny linoleum with the brightest of floral patterns; in a few of the wealthier homes the ben-end might have a carpet, but in any case there would be hearth rugs. These rugs were made of rags; every worn-out garment (and it would be worn-out after being handed down from parent to child) was cut into strips (about 1 inch by 3 inches) by the younger members of the family. These were then knitted in to form the pile of a rug with a base of thick strong string. I can remember in our own home my mother's rug, with its steel needles or pins, which must have been about six feet long. Balls of twine and bags of cut rags, after many long hours of hard work, were transformed into cosy hard-wearing rugs, soft and warm and almost everlasting. I can also remember how heavy they were when they had to be rolled up, dragged out to the 'backdoor', and there, if you were strong enough, shaken with such a sharp crack as to dislodge all the crumbs and minute pieces of debris which lodged in the close-woven fabric. When canvas backed rugs, worked with ready-cut lengths of thick wool, gradually grew in fashion, the old folks considered them not only extravagant but almost decadent by encouraging ease and laziness.

The walls were covered first by bright coloured and patterned wallpaper, varnished in the kitchen, and then, where there was space, by 'pictures' which, like the ornaments, changed with the changing fashions. Just before 1914, the fateful year which brought such changes into every department of life in the fishing community, it was essential to have coloured enlargements of family photographs, and so, above the fireplace in the ben-end, the stern and forbidding portraits of 'Dey' (grand-dad) and granny, frowned down on any convivial gathering.

Such then was the outward appearance of the homes; but homes do not consist of houses and furniture, and the home life of the fisher folk was as special as everything else. What struck me most as a child was that the fisher bairns addressed their mothers by their first names. In time I came to realise what was behind this lese-majesty or near-blasphemy as it appeared to me. Most of the fisher homes were 'owner-occupied' even at a time when renting of houses was far more usual. The head of the household and his family occupied the first floor – the 'up-the-stairs' – and when in the course of nature the old folks passed on, the eldest son and his family moved up, and his eldest son and family moved in to the ground floor. So there were usually three generations in what was practically the same household, for 'dey' and 'granny' would find accommodation in some corner when no longer able to lead independent lives. So of course the children heard their mothers called Maggie or Jean and naturally followed suit. As Christian names were given according to a strict pattern, it followed that there would be in the same family two or even three Maggies or Jeans, so they were distinguished by adding the father's Christian name – e.g., 'Tam's Maggie' or 'Dauvit's Jean'.

This custom of using the Christian name of any woman, married or single, was by no means confined to the family. 'Mrs' and 'Miss' were terms never used, even by the minister, and certainly never by the doctor. I remember on one occasion a girl in her early teens was 'going into service' with the wife of one of the better-off shopkeepers in Anstruther; her mother, in discussing her daughter's prospects said, in all seriousness, 'There's just a'e thing, she doesna ken the mistress's first name, so what will she call her?'

One story, whether apocryphal or true I know not, deserves a place here:

> A lady of fashion in London, while on a visit to one of the 'big houses' of the district, thought it would be a good idea to get one of these very attractive fisher girls and train her as a maid-servant. So she asked the local minister (an elderly bachelor who could have no idea of the practical problems involved) to find her a strong, willing girl who would be prepared to travel to London. In due course such an adventurous lassie was found and fitted out with the dozens of garments without which no self-respecting mother would allow a daughter to leave her house. Arrived in London, the girl was instructed in her duties by the head housemaid. Maggie – or Smith, as she must now be known – was to assist house and table. After clearing the breakfast table and collecting the breakfast trays, she was to don her black uniform dress with embroidered white apron, starched cuffs and cap with waist-long streamers, and be ready to answer the door and admit callers. By eleven o'clock Smith had finished all her tasks and never having known an hour of daylight when she had nothing to do, she looked about for some job which she could do to pass the time. The hall was of black and white tiles which to her eyes looked far from clean. So up to her attic bedroom she went and from her 'kist' she produced what was known as 'a coorse apron' – an ample garment made of strong hessian which covered her dainty uniform very

> effectually. Off came the stiff cuffs and the black sleeves were rolled up well above the elbow. The cap also had to be sacrificed as the streamers would interfere with operations. So equipped, she went off to the back premises and managed to secure a bucket of hot water, a bar of soap, a scrubbing brush and a washcloth. Down on her hands and knees and surrounded by a small sea of suds, she was scrubbing to her heart's content when she was interrupted by a peal on the front door bell. Just as she would have done at home, she got up from her knees, wiped the suds from her hands and arms on her apron and opened the front door. On the doorstep stood an elegant lady and gentleman who, though startled at the apparition, managed to ask if Lady Euphemia was at home. Smith said casually: 'I dinna ken. I'll gan and speir!' [I don't know. I'll go and ask]. Leaving the astounded callers on the doorstep, she went to the foot of the staircase and called out, in a voice which would have carried from pier to pier at home, 'Phemie, Phemie, are ye there Phemie?' It was some moments before it dawned on Lady Euphemia that she might be the one who was so hailed. In magnificent rage, she sailed down the stairs and began to ask Smith how she dared to behave so, but Smith, not a whit abashed, said: 'Och, never mind aboot that the noo. There's a man and a woman at the door wanting to see you.'

The story goes no further except to report that Maggie arrived home exactly a week after leaving, completely ignorant of the reason for her summary dismissal.

Poor Maggie had never been taught how to use leisure; I doubt whether she knew what leisure was until she left her fisher home. There women and girls were fully occupied from early morning till late at night. A recital of what the wife and mother of a normal family of six or seven children got through in the course of a day sounds completely incredible to a woman or girl of the present day, even though she may be one who is doing a 'full-time' job and running a home as well. In my young days, in any working-class home, the working day was at least sixteen hours long, from 6 a.m. to 10 p.m., and to most women that was scarcely long enough to do all that had to be done. But the fisher woman had many more duties to fit into the day than other women. She had to mend the nets, and though in a slovenly household clothes might be worn in holes, nets must be mended, and during the fishings from the home ports, this mending had to be fitted into the regular household routine. Every female in the household helped with this priority job. Even quite small girls could fill the wooden needles with fine cord and the older girls would have to mend the easy holes where only one or two meshes were involved, while the older women tackled the badly torn places where long experience and skill were needed to make a mend which would not stand out against the almost invisible net when it was in the water. In some households there were not enough women for the job, and so unattached spinsters were called in – unmarried daughters of older men who had left the fishing.

In addition to the net mending, the fisher women had to keep the men of the family supplied with all the knitted garments which were so essential. There was the navy-blue 'guernsey', knitted in elaborate patterns; sea-boot stockings, thigh-length and knitted of oily thick yarns so as to be practically waterproof; ordinary socks or stockings to wear under these; long drawers reaching from waist to ankle, and vests with long sleeves. Little wonder that on the rare occasions when a fisher woman went for a walk she knitted as she walked. The knitting needles were of steel, very long and fine in comparison with the present-day plastic needles, for the 'cloth' had to close and firm enough to keep out wind, rain and sea-water. Round the waist the knitter wore a 'shield' or 'busk', a belt of strong leather, the front of which was stuffed with straw and pierced with holes into which the ends of the long needles fitted for safety and convenience, for when a woman was knitting she was almost always doing something else at the same time.

In 1939 I was spending a holiday on the Norfolk coast, and I had a young son to fit out for boarding-school with six pairs of long stockings. To get them done in time meant that I had to knit most of the day, while the children played on the beach. And when we went for the daily walk along the cliff top I naturally took my knitting with me. To most of the holidaymakers I suppose I must have seemed an eccentric, but one old fisherman stopped me and said: 'Its a long time since I saw anyone knitting as she walked. But when I was a young man, no woman went out without her knitting.' So I suppose fisher folk were the same all round the coasts.

The routine of the household depended on the tides. When the boats fished from home they went out by the afternoon or evening tide, fished all night, and came back with the morning tide. This meant that on no two days was the timetable alike, and the women had to be ready with a hot meal at any time of the day or night. My father was a fish-buyer, and as he had to be on the pier whenever the boats came in, we had this same uncertain programme. I can remember so well being wakened in the very early hours by the sound of running feet (the man or boy employed to keep a lookout for the first sign of the boats coming home) and a shout of 'Alec, Alec, the boats are coming in.' Then I would hear Mother get up and while Father washed and dressed she would have a hot plate of soup, a pot of strong tea, and bread and butter, all ready within minutes. And that was before there was any such thing as cooking by gas, and electricity was, in our district at least, something to play jokes with. In our home, and I expect in most others, the secret lay in a fire that was never allowed to go out. There were no 'all-night burning stoves', just a 'gathering coal' backed by 'coom' or dust, a pot of soup on one side, and a kettle, always on the boil, on the other.

In these circumstances it is not surprising that fisher women were not famous as cooks. Food was something that had to be cheap, easily prepared and nourishing. Breakfast was porridge and milk, strong tea, bread and syrup or treacle; both varieties, 'golden' or 'black', were sold from barrels at the grocer's shop. Mid-day dinner was 'kail' or Scotch broth, to give it its English name. It was made from 'boiling beef' (flank, brisket or 'hough'), pearl barley, all or any root vegetables, except potatoes, onions or leeks, and curly kail greens. While this was boiling at the side of the kitchen range, a large pot or 'goblet' of potatoes would be on the other hob, and, if the men were at home, a dumpling, made of flour, suet, sugar and currants, would be boiled in a cloth in the kail pan, and so there was a three-course meal ready at any time. Tea at six o'clock was the last meal of the day, and the only one which would be added to for the benefit of visitors. At this meal there was always 'kitchen', i.e., something solid and savoury. Mostly it would be fish – salt herring boiled, fresh herring fried, smoked haddock grilled, or salt cod, boiled and made into a fish pie with potato. But when a visitor was expected boiled ham would be bought from the grocer and the 'plain bread', augmented with 'small bread' (plain dough buns [cookies], scones and soft biscuits, which were a kind of semi-sweet soft roll) or even, if it was a very special occasion, by cake and shortbread.

The weekend catering of course was special. From dinner-time on Saturday till Monday morning everything was different. As Friday evening and Saturday morning were spent in 'cleaning the house', nothing was cooked on the kitchen range on Saturday. After a session of blacklead and bathbrick it shone in all the splendour of its polished black and burnished steel, for thirty-six hours of the week an adornment to the house, of use only as a setting for the blazing, glowing coal fire. Gas had begun to be used a supplementary fuel for cooking, and most homes had a gas ring or even a small grill which was used for the weekend cooking. Saturday's dinner consisted of meat pies – baker's pies – which were in the shops, hot and ready to eat from midday each Saturday. No porridge on Sunday morning; that was the time for fried ham, steak and sausages. If there was any midday meal on Sunday it was just a cup of tea and possibly bread and jam – 'a jeely piece' for the bairns. For Sunday tea there would be stewed steak and bread, followed by shortbread and sultana cake, or 'black bun' if it was near New Year time, but these items would be all 'shop bought'; there was little or no home-made cake or pastry.

The clothes of the fisher folk were distinctive. The men wore navy-blue 'guernseys' and bell-bottomed trousers, except on Sundays, when they suffered the torture of stiff high collar and tie and a conventional three-piece suit, usually navy-blue, though there would always be a black suit for funerals and Communion Sundays. The women were different from the other working-class women only in their headgear, for the fisher women wore shawls, except on Sundays, when they wore the most extravagant hats, decorated or 'trimmed' with masses of artificial flowers in the summer and loads of feathers, even whole stuffed birds, in the winter. The older women in my youth still wore bonnets when 'dressed' to go out and mutches (white cotton or linen caps) indoors. These mutches were the sign of a married woman, and my mother, married at eighteen, broke all the conventions and was regarded as 'a brazen besom' when she refused to wear a mutch. Twenty years later, only the women of an older generation than hers still wore them.

The rules about where and when to wear shawl or hat, guernsey, or collar and tie, were fixed and unalterable. A meeting in the kirk, on whatever day or time, required full dress, but if it were held in the vestry or hall, then only shawls and guernseys would be worn. A visit to the doctor at his surgery called for a hat, though one could attend a morning surgery in an emergency in a shawl, and so the doctor found that his Sabbath was sadly misnamed 'a day of rest'. Doctors then worked a seven-day week, and it was so convenient, when one was dressed up in any case for the kirk, to drop round at the doctor's to get something for the 'hoast' [cough] or 'stamach'.

What I think I envied most about the fisher girls I knew was the fact that they and all members of the fishing community had three complete 'rigs' in place of two (every day and Sunday) which was all we could ever hope for. The third outfit was 'the Saturday-night rig', usually the Sunday clothes in their first remove from the peak of fashion, later to descend to everyday wear.

Because the fishermen went to distant places to fish for six or eight weeks at a time, they had to have a much larger stock of clothing than working men in other ways of life; and what the men had perforce, the women also had for their own satisfaction. On the somewhat few occasions, before the First World War, when a fishermen brought home an 'outling' bride, her trousseau was one of the first things to come under criticism. 'She has naething but what's on her back and one shift!' That took a lot of living down.

The economics of the home perhaps do not belong in this chapter, but it may be said that the solid pieces of furniture, such as tables, chairs and dressers, were the concern of the husband, and when a new home was set up, his mother, who had prior to his marriage received all his earnings, provided the items. The bride was responsible for the beds, bedding and linen, all soft furnishings, china, pictures and rugs. And to provide these a 'bottom drawer' was started for every girl as soon as she was born, and was filled by presents from 'the Sooth'. Little girls of three or four years of age might have complete tea and dinner sets, and I remember one, who did not belong to our district, but came from the North to the Winter Herring, whose mother boasted to us that she already had a silver tea service 'laid by'; the child was then just over two years old.

4. The Fishings

THE EARLY YEARS of the twentieth century proved to be the end of an era, not only because of the disruption caused by the First World War, preceded by adjustments made necessary by the introduction of steam drifters and motor engines into the sailing boats, but also because the herring, for some unknown reason, became more scarce, or rather less plentiful, and the white fish too were more difficult to get. I have not the scientific knowledge to understand how far these different features of the change were inter-dependent; I can only recall what happened in one small corner of the fisher-world.

There always had been good and bad seasons, and I might have believed that the stories of phenomenal shoals of herring coming close enough inshore to be scooped out in baskets were only old wives' tales had I not lived through just such an event, which had tragic results. It was during the 'Winter Herring' in the early

1900s, although I cannot remember dates and have no means of verifying them. The fishing had been good from the start and the herring were shoaling close inshore. In the early hours of one morning the *Morning Star*, a sailing boat from Pittenweem, was fishing just to the west side of Anstruther harbour. When they started to haul the nets, it was clear that they had a good catch. As the nets came over the side, it seemed as if every mesh had its fish, shaken out into the hold while the empty net was left on the deck. Net after net came up, and the men, intent on the good haul, paid little attention to the growing pile in the hold. By this time it was daylight, and watchers on the pier were becoming aware of an unusual catch. But while they watched, no doubt speculating on the number of crans the lucky crew had caught, the boat, the men, the nets and the herring sank. Only one man escaped, and he left the fishing forever and became a ploughman. When I worked in the town clerk's office at Pittenweem some ten or fifteen years later, the '*Morning Star* Disaster Fund' was one of my responsibilities, and I paid out, week by week, the dole which the surviving widows received.

A more amusing tale was told of an earlier shoal, which came inshore on the east side of Anstruther harbour. As the fishing boats came in laden to the gunwales, boys and shopkeepers took out every yawl or rowing boat they could find in order to share in the haul, and the fishwives ran along the 'skellies' to scoop out baskets full of herring from the deep water at the end of the rocks. The minister of West Anstruther came out from his study and cast envious eyes on the riches to be had for the taking. He walked along, past the harbours and out on to the Green, where the old men were watching the busy scene and doubtless each one would be telling his tale of the great shoals of his day. The minister, himself an old man, accosted one of the fishermen, whose son told me the story many times, and asked if he knew where he could borrow or hire a boat. 'No, sir, there's nae boat to be had the day.' And as the minister turned away, he called him back: 'I was reading in an old book the ither day, the story of two fishermen who were mending their nets when the Maister called them to be fishers of men. It said they left all and followed Him, so they didna take their boat with them. Mebbe ye could get a len' o'd!'

The fishing year began as soon as men had sobered up from Hogmanay and New Year's Day, with the 'Winter Herring', which was fished from home. In the weeks before Christmas the buyers started to get ready, and all round the harbour the stances began to fill up with boxes and barrels, and the coopers got busy, each with his little fire to heat the hoops. It was a wonderful playground for children, making 'houses' out of boxes set on end (that was a girls' game), while the boys chased each other along ranks of barrels or tried to ride a rolling barrel from one end of the stance to the other.

It was to this fishing that the North boats came, and as the wives and families came too, we had playmates with unfamiliar accents, and it was all the rage to 'speak North'. The boats too were different; one fleet was 'clinker' built and the other 'carvel', but which was the home fleet and which the other I never could make out; all I knew was that the stern of one lot went down straight while the stern of the other lot was curved in to the keel, and ours were the straight ones. They had different ways too of dealing with the catch. Our boats shook out the nets as they hauled them, and by the time they reached the harbour head the skipper could tell to within a basketful how many crans he had to sell. But the North men hauled the nets and came straight in to the harbour, and at their berth they emptied the nets. As a child I don't think I was especially perceptive and rarely noticed any particular beauty in everyday things, but I can recall now the sheer joy of running down to the harbour, on my way to the baker's shop, on a sunny frosty morning and seeing the boats drawn up and in each one the fishermen holding up the nets, full of silver herring, and shaking them into the hold. I don't think I have ever seen a more completely satisfying picture, and I wish I had an artist's fingers so that I could make it live again.

This fishing was the most important one of the year, if not for the fisher folk, at least for the tradespeople of the district. Fish-buyers came from England, for whom board and lodgings had to be provided; extra staff set up an additional telegraph office, and the goods station, for the rest of the year an empty platform except when an occasional coal waggon or two came in, became really alive, fully staffed with clerks and porters, prepared to work the clock round if necessary to clear the catch from one night's fishing before the next day's came in. The piers would be lined with carts and lorries, all horse-drawn, and as soon as the fish were landed, packed into boxes with salt, the carts were loaded, and at a full gallop they set off for the goods station.

'Where's there muck there's money' was as true in our

area as it was in industrial Yorkshire or Lancashire or wherever the phrase was born. But it was the fisher folk who got the money and the tradesmen and 'gentry' of the Anstruthers who got the muck. The piers and streets round the harbour were paved with causeway blocks and the ice, salt and scales spilt on them were easily washed into the gutter. But the main road from east to west was a 'metal' road, in the days before tar or asphalt were used for road-making. These roads were made by rolling broken whinstone [road metal] into soft earth or grit rolled in to fill up all the interstices. It was along this road, through Anstruther Easter and Anstruther Wester, that the lorries and carts were driven to the goods station, dripping the horrible mixture of salt, ice and fish scales to mix with the already muddy road surface, so that it became a sea of mud from one side to the other. Even the pavements were covered with this slime, and to cross the road one had to step into the gutters ankle-deep. The roadmen were busy all day long trying to keep crossing places clear, but their job was rather like Hercules' attempt to clean out the Augean stables.

Nevertheless it was a sad day for the tradesmen, as well as for the fishermen, when the roads were clear and clean, for if money did not come in through the harbour mouth it did not come in at all.

This fishing lasted for about ten weeks, and by the middle of March the herring nets were 'barked', dried and stored, and the boats started the 'gartlins' or greatlines, i.e., fishing with lines for white fish. 'The lines' was line fishing carried on all the year round by small boats, known locally as 'yawls', with a crew of four men, and the lines they used were of fine cord, baited with mussels. Anstruther, the port for Cellardyke, was not much used by line fishermen, though some of the older fishermen carried it on from Cellardyke harbour after they had retired from the herring fishing. The chief centre for line fishing was Pittenweem. There the women had an even harder life in some ways than the Cellardyke or St Monance women, for it was the job of the womenfolk to sort and bait the lines between tides. This work was done in the cellar, a stone-floored room at ground level with a door alongside the house door. There, in summer or in winter, in fair weather or foul, they sat 'redding' the tangled lines and baiting each one of hundreds of hooks with a mussel taken from a bucket standing beside each woman and shelled with a special knife. These mussels were brought daily by lorry from beds in the Tay estuary. As the hooks were baited they were laid evenly along the side of an oval-shaped shallow basket, layer upon layer, until the line was finished. This was a skilled occupation, as the lines had to run out free, and one hook laid askew could foul a line while it was being 'shot'.

The 'gartlins' was quite a different type of fishing, and for the weeks it lasted the women whose men were engaged in it had a somewhat easier life, or perhaps 'less pressing duties' would be a more apt description, for I have no doubt that every day was used to the full in mending old nets or 'mounting' new ones. This 'mounting' of nets was sometimes done by women who were employed at home on piece-rates by the shops who outfitted the boats. The nets came from the local factory simply as portions of netting of a certain length, width and mesh, and before they could be used in fishing a length of somewhat stronger netting about one foot wide had to be joined along one of the long sides. This stronger netting was fitted with lengths of string, 'nossels' I think these were called, to which could be attached the 'bladders' which kept the net upright in the water.

But to get back to this greatline fishing. I don't think I ever saw one of these lines, but I have seen the cod hooks many times. Instead of mussel bait, these large hooks were baited with herring (by this time spent and useless for the market), and the boats carried one or two nets to catch the bait before the fishing – it might be three nights or a week – began. The lines were 'shot' and hauled by a donkey engine, not by hand. When the boats came in from the sea, usually on a Saturday morning, the catch was landed and laid out on the pier. I remember when the piers were so closely covered with cod, ling, skate, halibut and turbot that there was scarcely room to put a foot down. Most of these fish would be sold fresh, packed in ice and despatched to the big markets, even as far as Billingsgate; but, it may have been when there was a glut, the cod were sometimes salted and dried on 'flakes' – long tables of planks on trestles. There the split fish had to be turned so that each side should be dried evenly and covered with tarpaulins if it should rain. This was a most laborious business, and it is much better managed by the Norwegian fishermen, who dry their cod on upright frames with both sides of the fish exposed to sun and air and protected from rain by easily drawn curtains.

After the greatline fishing the boats went back to herring fishing in the Drave or Summer Herring, which was, for the most part, a home-port fishing, though, if herring were scarce, some boats would go north, and some through the Caledonian Canal to the west coast. The Drave was the most exciting fishing of all for the children, for it coincided with the long summer holidays from school – nine weeks in which to haunt the harbour and piers. Once again the open space along the shore side of the harbours was occupied by boxes and barrels belonging to the various fish-buyers, but there was added interest for us in the presence of the 'crews' of women in the curing yards. I remember a number of years when, presumably because the fishing was very good, the buyers brought women from the north and west coasts, and we heard our first foreign language – Gaelic. The women worked in crews of three – two gutters to one packer. The herring were emptied into deep troughs, one to each crew, and each woman had ranged on the side of the trough opposite to her baskets or boxes to receive the various grades of herring, for the women not only gutted but graded the herring, and that at a quite incredible speed. I remember one of the summer visitors, who regarded this as one of the 'sights', timing a woman with a stop-watch (itself a curiosity to us), and she did sixty in one minute. There were four grades of herring: full, matty, spent and 't.b.' ('torn bellies', which means any damaged herring of any size or condition). The women used a very sharp, short knife and cut fingers were common. Because of the nature of the work, the slightest cut could quickly become infected and very painful, so the hands were protected by rag bandages and we children used to beg rags and hang around ready to render first-aid when it was needed. The third member of the crew collected the full baskets and packed them between layers of salt in barrels. If a packer was of short stature she nearly disappeared into the barrel as she packed the first few layers, and she would emerge bit by bit as the barrel filled up. When it came to within a few layers of the top, the children would swarm around and compete for the job of finishing it off. Packing was as skilled a job as gutting, and we used to watch, fascinated by the quickly moving fingers which just seemed to flicker over the heap of herring to transform them into neat layers, each layer crosswise to the one underneath.

As the barrels were filled the coopers took them away, fitted on the lid and hammered on the top hoop. They were then laid on their sides, each quality of herring by itself. After the end of the season, the tops were removed and the barrels where the herring had settled down as the salt dissolved to brine were topped from barrels of similar quality left for that purpose. Then brine was poured in, the tops were replaced, and the barrels were ready to be branded by the fishery officer. Sometimes he would ask for a barrel to be opened, but usually the brand was applied without question, and the barrels were ready for export to the European markets.

Another feature of the Drave was the sale of some herring as fresh herring to dealers or ‘cadgers’ who bought at the pierhead and sold the herring round the country. The boats came in as the tide suited, and that might be as early as 2 a.m. or as late as noon, so the cadgers, and their dogs, took up their positions when the pubs closed, sleeping in their carts, while the dogs kept watch. On the pier, around high tide, would be the fisher lads acting as volunteer look-outs. Long before daylight their keen eyes would pick out the boats heading for the harbour and they would call out the names. Before the boats were within hailing distance the buyers would be clustered round the pier-head and soon the estimated catches would be shouted by the skipper and the buyers would shout back their bids, until the day's orders for fresh herring for the English markets and herring for curing were filled. Sometimes the boats would have missed a tide and then the herrings were 'over-days' and sold at a lower price. When the buyers had taken the cream of the catch, the cadgers, who only wanted a box or two, took the rest.

When the catch was unloaded, the fisher boys were there, getting in the way of men and horses. The herring came ashore in baskets, slung by a crane, and there were four baskets to a crane, two to a box. As the baskets were tilted over the carts some herring were sure to fall, and on these the boys pounced. Out of their pockets would come cord or string, to which the herring were threaded by the gills, and off the boys would go to dispose of their catch. If the tide was an early one, the goodwives of Anstruther could buy a dozen herring, not ten hours out of the sea, and for a penny have a hearty breakfast for their families.

There was also the kippering trade, catered for by large kippering sheds. The work of splitting the herring, salting and threading them on splines, to be hung in the smoke of oak chips, was, I suppose, more skilled than that of the curing crews, but I don't remember ever seeing the women at work. But the smell of kippering herring is one that still comes back to me, possibly because the last of the large kippering sheds went up in smoke one winter's evening and smouldered for several days before the fire could be put out.

Among the crowds on the pier would be a few visitors, and anyone who was free to see the boats come in; these onlookers as often as not were presented with a string of herring. This custom of handing a string of herring or a fry of haddock or a cod to any friend or neighbour survived until the Second World War, when it was effectually killed by food control 'snoopers' who lurked around the piers to see that every fish landed went into the auction salesroom. I was paying a visit to Anstruther one winter in the later 1930s, and my small son, then aged eight or nine, had been trying his hand at catching 'dergies' or 'potlies' from the West Pier. I was dragging him home, empty-handed and terribly disappointed, when we passed a boat which had just landed its catch, and from which the crew were about to come ashore. On the deck there was a fish-box with several large cod, caught by the men for their own use, as they had been fishing for herring. It was a North boat and the men were all complete strangers to me, but as I passed they could not help overhearing Pat's clear voice: 'Look, Mummy, they've got a lot of fish. Could you not ask them for one?' While I was hush-hushing him one of the crew called up: 'We need them for oursel's, laddie', but the skipper overheard and said at once: 'Gie the bairn a fish, man.' And they loaded me up with the biggest of the catch. I had no means of conveying it home except by sticking my fingers through the gills and thus, with the fish at arm's length, lest my coat should come in contact with it, I made my painful way home, not to the wynd near the harbour where I had been brought up, but right through the Anstruther shopping district and out to fashionable West Anstruther, with an excited small boy tugging at my other arm and wondering, very audibly, if people would think he had caught it.

The last fishing of the year was 'the Sooth', fished from Yarmouth and Lowestoft. As the men were away from home for eight to ten weeks, their womenfolk went too, by special fish-workers' trains, as well as crews of fish-workers to cure the herring. For those left at home, the winter months were dull, but first of all there was all the excitement of their going. For this fishing the bakers made 'boat biscuit' which took the place of bread when the boats were at sea. As each boat left its moorings and started its long voyage, some of these biscuits were handed out to the waiting crowds of children. I often wondered why this special delicacy was not on sale all the year round, for they really were delicious, but on looking back I wonder whether some of the flavour came from the excitement and emotion of seeing the boats put out to sea for the longest voyage and absence of the year.

For the tradespeople and those who, like ourselves, were in but not of the fishing community, these winter months were dull indeed, but the fisher bairns had more excitement. Some of the smaller fry would be 'Sooth' with their mothers; the older ones, whose schooling was more important, might be looked after by a granny or an auntie, and these are notoriously fair game. News of the catches came by telegram to the fish-salesman who managed the boat, and a boy, usually the skipper's son, took the tidings round to the homes of the members of the crew – quite a lucrative errand if the catch was a good one.

When the boats came home, the town sprang into life

again. The harbour was full, and carts were busy taking nets and gear to the homes. The nets had to be 'barked' and then dried, and soon every grass field and brae round the town would be spread with the drying nets. Bedding and clothes had to be washed and backyards and bleaching greens were full.

For the children, of course, the chief interest was in what Father had brought from the Sooth. The traditional things were apples and oranges, nuts and grapes, 'rock' and 'sweeties' of every kind. Children of a lesser breed, such as my family, benefited from the overflow, and for perhaps a week we all had enough and to spare of the good things of life.

For the womenfolk this was the time when the china cupboard was replenished and the bottom drawers filled. Occasionally there would be a bride from the South whose trousseau or 'providing' looked meagre indeed as everyone watched to see how many 'kists', or chests, it filled. Her lot was not to be envied, for to the end of her days she would be 'English' and therefore unable to measure up to the standard of a fisherman's wife.

Indeed some had 'English' added as an extra Christian name, and I can remember old women who were still so named, though their children and grandchildren had been born and brought up as Scots as any.

And so ended the Fishing Year, with Christmas scarcely acknowledged as a festival and Hogmanay and New Year offering a brief cessation of toil and the only occasion of the year when one could be idle without feeling guilty, if there were not far more exciting things to fill the time.

Before leaving the subject of the fishings, perhaps I

ought to mention one which was of particular interest to the children. This was 'wulk' [winkle] gathering which was a source of pocket money to all the fisher boys who wanted to earn an honest penny; I don't think the girls shared the spoil, though they might be allowed to share the toil. Behind the East Pier and the 'blocks' stretched the 'skellies' – shelves of limestone or sandstone running out from the shore to the deep sea beyond the lowest tide line, titled at an angle of 30° or so and separated from each other by valleys strewn with seaweed-covered rocks. On these and on the broken side of the skellies there were winkles by the thousand, and anyone at low tide could fill a pail without walking far, though of course the biggest specimens were found farthest from the shore, where only the big boys with long legs could venture through the pools.

The buyer for this fishing was an old woman who was one of the real characters of the place. She was known as 'Pie Liz', but the origin of the nickname is unknown to me. She lived up a close near my grandmother's old house, and I often saw the boys bring their catches to her. It was no use trying to palm off small ones at the bottom of the pail, for she poured the catch into a sieve, and those winkles which did not make the grade were thrown out. Pie Liz had, I believe, some arrangement with one of the buyers who dealt in crabs and lobsters – another small branch of the fishings, carried on mostly by old men after retiring from deep-sea fishing. In due course the 'wulks' gathered on the Cellardyke skellies were sold as winkles at Billingsgate.

Pie Liz was one of the most 'postit' women of the town. She knew every detail of a scandal or a romance as soon, or even sooner, than the parties to the case, and nothing escaped her notice. She also had a way of living by barter which she had brought to a fine art. She had a garden in which she grew, among other things, early potatoes, and it was her boast that her potatoes were earliest of all. This was little to wonder at, because she dug them at the stage when it was quite a triumph to be able to take two bites out of one of them. As soon as her potatoes were reasonably sized, she would dig up a small basketful and take them to a fisherman's wife, making sure beforehand that the boat was in harbour with a good catch. In return for the potatoes she would be given a string of herring or a fry of haddock or flounders, or a cod or ling. Enough of that haul would be left at home for her next meal, and the remainder would be cleaned and neatly packed with some more of her potatoes, and this time she would visit someone who kept hens in their backyard. After she had offered her present, the basket would be returned with perhaps half a dozen eggs. Three of these, offered to a housewife famous for her home baking, would be worth a few scones or bannocks and perhaps a pot of home-made jam as well. When, much later in life, I visited North Africa and was offered small presents of trifling value when making a casual call on an Arab household, only to be told *sotto voce* by a more experienced companion that I was expected to respond by giving a 'present' of greater value in exchange, I thought how much at home Pie Liz would have been among the Arabs of Algeria or the Egyptians of the Nile Delta.

5. The Harbour

THE HARBOUR WAS THE CENTRE OF LIFE, not only for the fisher folk but also for the tradespeople and even, in a lesser degree in the days of my childhood, for the farming community. Tramp steamers brought in cargoes of 'cake' for the cattle and took away potatoes, and as this traffic took place when the boats were 'Sooth' I suppose it was a help financially in running the harbour.

For me, personally, the harbour played a very large part in my existence. My maternal great-grandfather was a farm labourer, living and working some miles from the coast with probably no connection with the sea. But one day he was seized by the press-gang and carried off to serve in the Navy during the Napoleonic wars; he eventually died in a French prison. His wife was left with the normal brood of six to a dozen young children. Her husband's people seem to have been somewhat better off than she was, and when the news reached them of the plight in which she was left, a deputation went to see her to find out what the family could do to help. The desolate widow met them on the threshold of her cot-house, and, without waiting for any greeting, confronted them with: 'If you have to come to offer sympathy, you are welcome; if you have come to pity, you can go back as you came.' What they answered is not recorded, but, in the event, Ann Fleming, or Pringle, was left with her pride as her only asset. My grandfather was a boy of seven or eight, quite old enough to earn a few pence by watching the cows on the laird's land. The laird had a son of about the same age, who formed a true friendship with the little cowherd. He was being educated at home and brought his books into the field where he taught my grandfather all he had already learned and then shared his lessons with him.

In due course my grandfather married and took up the life of a ploughman. What economic or other urge drove him out of his country life I never heard, but Anstruther harbour was being enlarged and made safer by the building of a breakwater and my grandfather moved from the country and became a labourer on the harbour construction. How long he was content or compelled to remain a labourer I don't know, but he was an educated man by the standards of the time, and later in life he became Clerk to the Harbour Commissioners.

My father, though born of fisher parents in West Fife, had been put to work in the coal pits, where wages were sure if not large. He hated the life so much that at the age of sixteen he ran away from home, came to Anstruther, and somehow contrived to become apprenticed to a firm of fish-buyers as a cooper. So it was the harbour that brought my father and mother together, and, as we lived but a stone's throw from it, it was the focus of life for all their large family.

The harbour, as I remember it, consisted of the outer harbour, bounded by the whole length of the East Pier and the extreme part of the West Pier and the Middle Pier; the inner harbour, separated from the outer harbour by the Cut-Mouth where the Middle Pier approached the West Pier leaving a channel only wide enough to take one boat at a time. It was, and still is, so far as I know, a tidal harbour only, though the project of making it into a deep-water harbour had been local politics for all my lifetime and no doubt for years before I was born.

The administration of the harbour was in the hands of the Harbour Commissioners, appointed, I believe, from the Town Councils of the Anstruthers and Cellardyke. The representatives from the Anstruther Council, being mostly tradesmen, might know little or nothing of the sea or boats or even of the harbour as a working unit. It was reported that a local draper contributed to a discussion of the Deep Water Harbour Project the original suggestion that six feet might be added to the height of the piers! This same gentleman, in the days of sail, thought he could solve the problem of a becalmed fishing fleet by erecting a huge windmill on the May Island.

The East Pier runs from the Harbour Head straight out to sea for about a quarter of a mile, and then makes a right-angle turn of about 100 yards to the Harbour Mouth, the channel into the harbour from the open sea. This channel did not seem very wide to us as children, and it was our ambition to wade across it. Although at a very low tide there would be a strip of sand or mud

Belle Patrick's Pringle grandparents in old age

Belle Patrick's mother, Agnes Pringle, photographed with her family in about 1865. Agnes is on the left. Her mother is holding a photograph of her son Tom ('Oor Tam'), who had emigrated to South Africa, so that he could be included in the group. When Agnes was widowed in 1904, Tam sent the money from South Africa to buy 'Algoa' in Shore Road, West Anstruther, where she spent fifty-two years of widowhood.

Alec Patrick

The Patrick family in about 1901. *Back row*: Lizzie, Nettie, Chrissie, Jim (the eldest). *Front row*: Alec Patrick, Annie, Willie, Dave (in army uniform with John, the youngest, in front of him), Belle, Rita. If Agnes wanted John, whom she spoined dreadfully after his father's death, she would shout 'Annie–Willie–Bella–Rita–JOHN'. Picture taken at Ireland's Photographic Studio, Shore St, Anstruther.

Belle Patrick with two of her sisters, a little before the outbreak of the First World War

Three sisters by the Dreel Burn in about 1920: Rita, Annie and Belle

Belle Patrick

Belle's wedding,
Edinburgh, 1927

Belle Patrick's birthplace, Anstruther Easter

George Street, Cellardyke, in 1910

Alec Patrick with the fisher lassies,
filling the herring barrels, and his beloved dog

Anstruther Harbour in about 1900
(courtesy of the Scottish Fisheries Museum)

The steam drifter *Noontide* KY 163 (skipper: James Brunton), Anstruther (courtesy of the Scottish Fisheries Museum)

Anstruther West Pier in about 1900 (courtesy of the Scottish Fisheries Museum)

Arrival of lifeboat *James and Mary Walker*, 30 July 1904
(courtesy of the Scottish Fisheries Museum)

News of the accession of George V, 1910
(courtesy of the Scottish Fisheries Museum)

exposed at the end of each pier, the channel between was still quite deep and I never knew anyone who managed it. The West Pier is longer than the East Pier and runs west from the Harbour Mouth to the Middle Pier, which bounds the outer harbour on the west, curving round from there to enclose the inner harbour and finishing at the west end of Shore Street.

The East Pier is built of large blocks of dressed stone and is some twelve feet wide, with a high wall protecting it from the sea. On the other side of the wall, which is wide enough to walk on, is the breakwater or 'blocks' as we called them. Each name is expressive, the first of the purpose of the barrier, the second of its construction: large blocks of concrete which looked as if some giant child had thrown them against the harbour wall when he tired of his game. The east–west part of the pier is surfaced with concrete and has a landing stage halfway down the inner side, reached by steps. This was a favourite place for the boys to fish for dergies or potlies when the tide was suitable; as the tide rose higher they would retreat to the pier proper. The 'halflins', as teenagers were called, disdained such safe and easy sport. They climbed down on to the blocks and fished there for codling or saithe. The smaller boys used a similar landing stage on the side of the Middle Pier and there they often caught flounders – 'flukes'. Even the girls tried their hand at fishing there, using sewing cotton and bent pins, but I, for one, never caught a thing. At the end of the East Pier stood the harbour light, housed in a cast-iron cylindrical tower, reached by footholds made in the structure. From this tower a green or red light could be shown to indicate the state of the tide, whether the Harbour Mouth was navigable or not. At the shore end of the Pier stood another light in a wooden frame, and I suppose these two lights were aligned so as to indicate the channel.

The Middle Pier was scarcely a pier at all in our eyes. It was just a wide quay dividing the two harbours. At the shore end was the lifeboat, a very primitive affair as I remember it first, built of cork – or so we children believed – so that it could not sink, and parked on the quay under the shelter of a tarpaulin. But while I was still a small child the lifeboat-house was built, just above the slipway which was the means of getting the lifeboat into the harbour. When the house was ready the new lifeboat arrived, escorted from the railway station by all the children of the neighbourhood. It was the most beautiful boat we had ever seen, and the great occasion of opening the new lifeboat-house and launching the new boat is one that I can never forget. I must leave it to others better qualified to describe the succession of lifeboats which followed the old 'cork' one, each better equipped than the last, but the name 'Lifeboat' brings back to me the glories of that day, and the spectacle of the blue, red and white boat with its crew in immaculate gear making its first appearance in the harbour. Only on the annual flag-day for the RNLI was there a repeat performance for the benefit of the summer visitors; when the lifeboat went to sea in real earnest, it was not on days of brilliant sunshine to the cheers of excited children and the music of the local band. The saga of these expeditions is worthy of a book of its own.

The other feature of the Middle Pier which I remember was the mounds of ballast, round stones about the size of a grapefruit, which came, I believe, from the May Island.

It was behind these ballast heaps and in the corrugated-iron open shed at the end of the pier that we used to see the young fishermen playing cards.

When I knew it first the West Pier closely resembled the East Pier in construction, except that the shore end of harbour wall was lower and broader and had no breakwater, because only the estuary of the Dreel Burn – the West sands – lay behind it, and not the open sea. As the pier curved round to the straight stretch at the end, the shelter of the estuary was lost, and consequently the wall was higher and was backed by a breakwater to the Cut Mouth, where the Inner Harbour was left behind. The last part of the pier, bounding the outer harbour, was a narrow concrete quay with only an open fence of iron posts and rails on the sea side, and a breakwater of small blocks of concrete several feet below the level of the pier. At the end of the West Pier was the handsome lighthouse known as 'The Hannah Harvey', after the donor of the concrete tower with an internal staircase. It was only when I was older that I knew why the lighthouse was so named. My grandfather was a rhymester, and among his poems which came into my mother's possession was one written as a tribute to Hannah Harvey in gratitude for her munificent gift. What we appreciated most about the Hannah Harvey was the smooth concrete bench built into the base. The building is sexagonal so there was usually one side sheltered from the wind where we girls could play our games of 'chuckles' and the older folk could rest and enjoy the view. However this pier was quite impassable beyond the end of the harbour wall when the weather was rough, as at high tide it was waveswept by the swell as well as by the breaking waves.

The East Pier, on the other hand, was passable, if not particularly safe, in the shelter of its high wall, though in a storm the breaking waves would run in a flurry of spray the whole length of the pier, reaching even the sea wall facing East Green by the gasworks. I remember getting completely soaked as I walked to Cellardyke Church for a special children's service one Sunday afternoon. I always loved the breaking waves, and I chose the shore road to Cellardyke so that I could watch them. It must have been the fabled seventh wave which broke in a magnificent arch of spray, reaching over about twenty feet of roadway to fall on my hapless head as I crouched against the gasworks wall.

The inner harbour silted up with mud, and each year, when the boats had left for the South, the dredger paid a visit. How we loved to watch the endless chain of buckets coming up full of mud – glorious mud! – to cascade into the open barges which were afterwards towed out to sea and emptied of their cargo, which, I suppose, immediately began its shoreward journey once more.

In the early years of the century it was decided to widen the West Pier on the inner side, and for this work gangs of Irish navvies were imported. Of the details of the work I remember nothing, and I don't suppose I would have remembered this widening of the pier had it not been for one frightening episode which made a lasting impression on me. Some kind of steam engine was in use on the pier and we children were playing on the Middle Pier, taking little notice of what was happening across the harbour, when the boiler of the engine burst with a great explosion and clouds of steam. One of the navvies had either been on the engine or working very near to it, and in a moment he was catapulted from the pier head-first into the mud of the harbour bottom. His rescue was effected somehow, to save him from a horrible death, but the sight of his boots sticking up out of the mud changed what had been a dirty but delightful plaything into a terrifying danger lurking for unwary feet such as mine. Never again would I dabble in mud to see how far I could go without falling in, and even now when I sing the metrical version of Psalm 40 the picture which flashes into my mind is not of an Eastern clay pit but of Anstruther harbour mud and the rescued man at length safe on the pier:

> He took me from a fearful pit,
> and from the miry clay,
> And on a rock he set my feet,
> establishing my way.

As I remember the harbour as a child, it was a very busy place indeed, and children were not particularly welcome on the piers. The harbour-master was always an ex-fisherman, ashore because of some physical infirmity, and the one I remember best was a very excitable man. As the boats raced in, always seemingly in danger of running each other down, he would dance up and down shouting instructions for berthing above the clamour of bids, seagulls and boys. Once when a boat was slow in approaching the pier after the line had been thrown round the bollard, he was heard to shout: 'Haul that pier tae to the vessel, you —— —— ——.' There was no censorship of language on the pier.

Not only fishing boats used the harbour in those far-off days, but ships carrying supplies for the fishing industry – barrel staves and other timber, salt and ice – and there was an outward trade too of cured herring, potatoes and other agricultural produce. Not all the ships were British and we children used to hang about near enough to hear the foreign sailors talking in their own lingo. The ice-ships attracted us too, because we could always pick up splinters of ice when they were being unloaded. Ice-lollies were things of the future, but I think we got as much pleasure out of the slithers of ice melting in our sticky, not-too-clean fingers as the youngsters of today get from their hygienic dollops of frozen fruit-flavoured water nibbled off a stick. In summer too we had occasional visits from pleasure steamers, making a call *en route* from the seaside resorts farther up the Firth to May Island, the Bass Rock and North Berwick. These gave us glimpses of high fashion as the 'visitors' came ashore in the glory of their holiday dress. We denizens of the harbour were dressed in our oldest and shabbiest attire, for school clothes were carefully put away at the beginning of the holidays, and boots and stockings were only worn on Sundays. On one occasion I was so filled with envy at seeing little girls coming ashore dressed in spotless white dresses that I determined to be equal with them. So I ran at full speed from the pier to our house and in some way or another persuaded mother to let me wear my one and only white muslin dress and the ankle-bar patent shoes which I had been given for school closing day. By the time I reached the pier, the visitors had dispersed and there was no one to be impressed by my finery. However, some of my special pals were making a fire on the beach to boil some winkles, and, by the time I had finished helping with the cooking and eating, my shoes – the only pair I possessed – were completely ruined, and my muslin frills were scarcely distinguishable from the drab clothes of my companions. *Sic transit gloria mundi.*

So much for the harbour as a place of business. For us children it was far more than that. It was playground, club and workshop, and schoolroom, where we learned more of life than in the orthodox classes.

The outer harbour was the playground, for the inner harbour was feet deep in mud and had no sloping beach. Our harbour had a sandy bottom, or rather a series of sandbanks separated by deep, muddy or rocky valleys, where there were always pools of water, even at spring tides, when the sea retreated beyond the end of the piers. These sandbanks were wonderful places to play when the tide was out, and part of the fun was to defer retreat to the beach before the flowing tide until the last possible moment. In keeping with the fashion of the day, we wore clothes which covered us completely, except for our legs and feet, which were always bare. But as the summer days got warmer, we would discard, one by one, the various layers of underclothes, until most of us wore only the long-sleeved, high-necked serge dress which was our uniform. Not even the smallest child would have bathed unless clad in a 'bathing costume', but wading was another matter. So long as we had our frocks on our shoulders we were modestly dressed, and as the water got deeper, so our frocks were hitched higher, until they were bunched round our armpits, but they could be lowered gradually as the water got more shallow and so we usually came safely to shore. While the tide was out there was lots to do, with sandbanks as our base. There were wulks and limpets to gather from the rocks and in the muddy creeks we could wade in the warm water and feel the baby flounders tickling the soles of our feet. Sometimes we would catch them, but mostly we left them alone; they were no use for eating.

At the west side of the harbour, just at the corner made by the Middle Pier, was the boatbuilders' yard. We were not very welcome there when the men were working, but as soon as they left we took possession. It was not all fun and games in the boatbuilders' yard, for there was kindling to gather from under and around the workmen's benches. But the timber, spread on long racks for weathering, made wonderful houses or shops, and we could always improvise a seesaw from loose planks or a swing from ropes left dangling from the partly-built boat on the stocks. As the ice cargoes made up to us for the lack of ice-lollies, so the pitch boilers provided us with a substitute for chewing gum, and without any thought of the therapeutic value of the product, we broke off the bubbles of soft tar from the edges of the boilers and chewed away to our heart's content. We believed, with what justification I know not, that this was very good for our teeth.

The yard also furnished us with materials for our beach fires, where we boiled our wulks in sea-water in rusty tins, and, even more unhygienically, fried limpets on any flat piece of iron we could find. Wulks were ready when the water had boiled over three times, and the limpets when they could be scraped off the home-made skillets. Both delicacies were eaten as soon as they were cool enough, and I remember they tasted chiefly of sand, which gritted our teeth as we ate them.

On the sloping sandy beach we made sand-pies, built castles, defied the tide with harbour-works and did all the things usually done at the seaside. Of course we did not often possess pails and spades, but the ubiquitous tins and driftwood made admirable substitutes. There was a stretch of sand which was above all but the highest tides, and there, one memorable summer, the older children built a hotel. I believe the enterprise was started when someone found a cooking stove which had been thrown out of a boat and the girls discovered that they could really use it. So the boys were called in to dig a hole deep enough to hold the stove and wide enough to let the girls work around it. Having finished the kitchen, there was nothing more for the boys to do, so they set to and excavated a series of rooms, furnished from the flotsam and jetsam readily available on the beach. The little fry were not allowed within those precincts; our job was to procure anything necessary for furnishing, cleaning and cooking which could not be found nearby. The stove had to be polished with blacklead before any self-respecting girl could use it, so one of us had to plunder the cleaning box at home. One girl wanted to make jam, and we were despatched to the nearest greengrocer's shop to forage in the refuse bins for over-ripe plums, and we each had to save, or steal, some sugar at each meal-time. Teas, of course, had to be supplied to the workmen, who became guests as soon as anything edible was produced, so tea and sugar had to come from somewhere. My eldest brother happened to be at home from Edinburgh on holiday at the height of this hotel season. He was sometimes on night duty in the telegraph office where he worked in the city, and he had discovered compressed tea tablets and saccharine tablets, both novelties then. I don't know if he ever found out what happened to his supplies after he had shown them to us, but they made good tea and were very popular at the 'hotel'.

Though Anstruther was developing as a seaside resort, not many visitors came to the East Harbour, except at Glasgow Fair time, and they were mostly of the poorer classes who, it was said, pawned their blankets to pay their fare. The gentry or toffs from Edinburgh were mostly accommodated in West Anstruther, where the Billowness with a few bathing huts and a nine-hole golf-course seemed to us Eastenders the height of sophistication. But even the children of the Glasgow 'keelies' had special holiday outfits, and what we envied most was the rubber plimsolls which they all wore on the beach. Like children everywhere, what we envied and coveted, we ridiculed, and when the tide was right we would go out to the sand banks and on the far side step into the mud until our feet were thickly coated to the ankle. Then, carefully avoiding the water, we would regain the beach and parade before the Glasgow children in our 'plimsolls', mimicking the way they drew back from the waves and shouting in Glasgow accents: 'Dinna get your shune weet, or yer mither'll skelp ye.' Sometimes the game went further and we plastered our legs with mud so that we could have 'wellingtons', and it usually finished up with one of us falling flat in the mud and having to go home and face the music. For me, these games ceased after the burst boiler episode; I was too scared that by some means or other the mud would get me.

Of course when the boats were coming in we left the beach to take part in all the excitement of the piers, and what happened there I have described in an earlier chapter.

Since the days of my childhood much has been done to improve the harbour, but the fishing, as I knew it then, has gone. I cannot write of the harbour and the fishing as it is today or as it may be in a decade. Perhaps a future historian will add the next chapter to the story.

6. Industries

In the time of which I write, all the industries in the area were directly connected with the fishing industry, except for a small golf-cleek factory. First of all and most important was the fish-curing, to which I have referred while dealing with the fishing itself. Though on occasions curing stations were set up round the harbour, each firm of fish-curers had one or more large yards, enclosed by sheds open to the yard. On the outside wall, on the street side, there were hatches or boles which allowed the herring to be emptied directly from the lorries into the shallow troughs where the women gutters worked. One of the side walls had double doors to admit carts and lorries and on the far side the coopers had their fires and workshops. Children were not admitted, except on business, and my recollections, such as they are, arise from the fact that when Father had a busy day in the yard one or other of the children had to take him his tea and so gained admittance. It was in those yards that the herring were gutted, packed and cured in the ordinary course of business, and I believe it was only the overflow in an exceptionally heavy fishing which was dealt with at the harbour and gave us so much excitement and pleasure.

The white fish were also prepared for the market in the fish yards; all that was required was that they should be sorted into species, weight and grade and packed in boxes with layers of crushed ice. I do not know whether the drying process which I have already described was normal procedure or merely a method of dealing with a glut of cod; I rather imagine it was the latter.

There was a subsidiary industry arising out of the greatline fishing – the manufacture of cod liver oil. How important this branch of fish processing was I have no idea; I think it came to a standstill, or perhaps slowed down to a stop, when I was quite small, possibly about 1900. The memory which fixes this activity in my mind is a very personal one. One of my father's duties in the yards was to supervise the boiling-down of the cod livers which produced the oil, and so penetrating was the smell of boiling oil that, despite the fact that my father invariably washed thoroughly and changed his suit before coming to the table, we children knew the moment we opened the door that Father was at the oil-making. The present generation, having had the essential vitamins or proteins, or whatever it is which fish oil abounds in, administered in tasteless capsules of the more concentrated halibut liver oil, probably know nothing of the nauseating, clinging odour of cod liver oil. What Father brought in with him was the smell of the raw material, not the refined essence for human consumption. On those days I, for one, could never eat any dinner, and by some accidental coincidence I associated it in my mind with scotch broth 'kail', and became firmly convinced that I hated it. Potato soup, on the other hand, was lovely. In our home, as in the average household of the day, fads and fancies were not tolerated. What was set before you had to be eaten and no arguments or explanation were allowed. One Sunday the whole family came home from church ready for the dinner which had been cooking on the hob. I was as hungry as the rest of the brood, but when the soup came to the table my heart sank, or rather my stomach rose, and I knew I could not swallow the stuff; it was kail.

Our family was a large one and the table in the middle of the room accommodated Father and Mother and the 'big ones'; the 'wee ones' had to be content with stools in front of the wide window seat which served as a table. As I sat and looked at the revolting stuff I had a brilliant idea. It was kail I did not like; why not turn it into potato soup by the simple addition of a lovely mealy potato, already steaming on the table in readiness for the second course? No sooner thought than done, and soon I was mashing up the potato till the stuff in my plate remotely resembled potato soup. But by this time the mixture was lukewarm, and when I ladled a big spoonful into my mouth I immediately spat it out again; I simply could not swallow it. With our backs to the table, our behaviour was not observed by the adults unless we made a noise and drew attention to ourselves, so, quite quietly I lifted my plate and passed through to our bedroom, a little room over the front door and adjoining the kitchen where we had our meals. There, still quietly, I lifted the lower sash of the window and, spoonful by spoonful, I emptied the soup outside. I was still too young to wonder where it went; sufficient to know it had gone. I presented my empty plate to mother and received my portion of boiled beef and potatoes to which, and to the following dumpling, I did ample justice.

As usual we all prepared for afternoon service, and as it was a fine day and Father wanted to see how the garden was looking, we went out by the back door. After the service we walked home by the shore road and approached the house from the front. It had had a new coat of whitewash during the week and mother was remarking to father how fresh it looked. When we came up to the front doorstep, there, right in the centre, was a mess of kail and potato, and as mother lifted horrified eyes to look for some explanation, she saw the new whitewash streaked with grease from the window sill to the lintel. No wonder I still remember the smell of boiling cod livers!

Another unpleasant incident involving fish oil concerned a whale which became stranded and expired in or behind the harbour. As it proved impossible to remove the carcass, it was decided to cut up the blubber and render it into oil in the cod liver oil boilers. I heard the story so often that I almost believe I was present when the tragedy occurred. Everybody in the neighbourhood was interested in the whale and more than interested in getting rid of the carcass before the town became uninhabitable. As Father was in charge it was only natural that some of the children should parade their intimate connection with the business, and so it came about that my sister Annie was one of a band of children who called in at the yard for a strictly unofficial visit of inspection. Poor Annie, she was one of the world's victims, and if anything had to go wrong, Annie was sure to be involved. Fortunately she did not fall into the cauldron of boiling oil, but she tumbled into the sump from which the refuse was collected. Mother always said it took a week or more to get rid of the smell!

Along the coast salmon fishing had developed in rather a small way, but it had no connection with the fisher folk. It was run by a company based in one of the ports north of the Tay, which leased the fishing rights from the Crown or whoever it was who owned them. The salmon cobbles were flat-bottomed boats which did not use a harbour, but some estuary or bay with a shelving beach where the cobbles could be hauled ashore. The nets were of much larger mesh than herring nets and consisted of a trap, open towards the sea and suspended from a pole. From the side of this trap long 'leaders' stretched out some distance into the sea and formed a sort of lane which made the fish swim straight on into the trap at the shore end. The salmon fishing was seasonal, from May to September, and the salmon-fishers were not locals, except for the man in charge of the station, who lived in West Anstruther. The station I knew best was at the Billowness, or rather the Hynd, and the cobbles were upturned for the winter on a grass patch about the beach, which is now part of the golf course. They made wonderful slides when their smooth bottoms were covered with rime or ice. I believe there was another station at Caiplie Cove, between Cellardyke and Crail.

Boatbuilding was the next most important industry. The boatbuilders' yard was almost part of the harbour, lying at the west corner between the Outer Harbour and the Middle Pier. Just alongside the slipway used by the lifeboat was the launching slip where the newly-built boats took to the water. Construction was completed on scaffolding immediately above the launching slipway, the bow being supported by wedges which could be knocked out so that the boat tilted forward, ready to slide down the double concrete piers which, on the occasion of a launching, were well anointed with soft soap. That launching slipway was one of our favourite play places. The smooth concrete piers sloped gently from the level of the boatbuilders' yard to the bottom of the harbour. At high tide all but the top few feet were submerged, but at low tide the sea retreated to the harbour mouth and we had a double row of 'houses'. The supporting buttresses which divided the 'houses' were in the form of steps where we could arrange shells, coloured stones and broken crockery as 'ornaments', and the concrete floor of the 'houses', though rougher than the smooth top of the slip, was level enough to be 'caumed' with limestone like the doorsteps and scullery floors at home. After a launching we had enough soap to scrub our floors white even without the limestone covering.

Behind the stocks on which the boats were built was the yard, with its piles of seasoning timber, and there too we played if the men were busy on the boat. But when the stocks were empty, the men were all busy in the yard and we had to dodge or be sent off. First of all, there would be the keel – a straight piece of oak a foot square which may have been trimmed *in situ* on the stocks. I don't remember seeing men working on that part, but the side timbers were fashioned in the yard, and it was fascinating to watch the skilled men with their adzes producing the symmetrical curves which gave the boat its strength and beauty. But our interest was neither in the skill of the workmen nor the beauty of their work; what we wanted was the long slithers of wood which fell away with every smooth stroke, and the boldest among us risked the amputation of a hand to snatch them before they reached the ground. No wonder the men yelled at us and the foreman chased us off with curses, but before he had got rid of one group, another would be after the spoil. If we could not get 'sticks' from the men working on the timbers, we would pester those who were planing the planks. The curls which fell from their planes were useless for making fires, but they made wonderful blonde ringlets which transformed our straggly locks into the most elegant coiffures, when we wanted to play at fashionable ladies.

All the work in the boatbuilders' yard was hand work, except for the screeching saw which had a house of its own in one corner of the yard. Just beside it was the boiler where the pitch was boiled for caulking.

When the timbers had been bolted on to the keel and the planks nailed to the timbers then the seams had to be caulked. The caulkers sat on cradles with a great ball of tow and a tool with which to work it into the crevices between the planks. Then the seam was sealed with hot pitch, leaving a streaky black scar along the clean amber of the pine. It was the drippings of this pitch which we used for chewing gum, and our eagerness to pick up the globules before they were too hard usually resulted in getting tar all over our hands and legs. Then Mother had to have a session with the grease pot till we were clean again.

While she was scrubbing away with the grease, Mother would tell us the story of the two women, one from the North where 'wh' was pronounced 'ph'. The North woman said to Fifer: 'Phat tak's aff tar?' The Fifer replied: 'Ay, fat tak's aff tar.' 'But phat tak's aff tar?' 'Ay, ay, fat tak's aff tar.' Mother would be the stupid North woman and we the clever Fifer, and the conversation would be repeated over and over again until the last spot was removed and off we would go to get another dab or two.

The yard was supposed to be locked up when the men stopped work, but that was only when it was too dark to see, and even if the gates on the road were bolted and barred we could always get in from the sea side. What need for an amusement park when a plank over a pile of wood made an ideal seesaw and a rope left dangling from the scaffolding round the boat could be easily looped up to make a swing? Though not aware of it at the time, I have no doubt, on looking back, that we did no end of damage, and under present conditions would have been labelled 'young delinquents' and brought to Court. As it was, the unlucky ones got a cuff on the head or a kick in the pants and learned to dodge next time.

The launchings were great occasions and worthy of Sunday clothes, if we could persuade mother of the importance of being in the front. If not, then we had to scramble up the 'Battery' as we called the curved retaining wall which separated the yard from the harbour – a dangerous proceeding at high-tide – and perch along the top. But from that position the boat was between us and the principles in the name-giving ceremony, and we could not even hear the smash of the bottle. But at least we had a clearer view of the men, with hammers poised, who knocked out the wedges, and we were near enough to see the first shudder of movement as the boat moved to the slip and then into the water. It was a wonderful thrill, but for me, at least, the occasion ended there. Often I would know the girl who named the boat, the daughter of the skipper as a rule, but I was never close enough in friendship to be included in the subsequent celebrations. I was an 'outsider', but I don't remember even wondering what happened when the crowd went home. Nowadays, of course, one is informed that such experiences build up tensions and frustrations which manifest themselves in later life, but then neither children nor adults were possessed of such knowledge, and I believe that it was a case of ignorance being bliss. Of course this book may be my reaction, somewhat delayed perhaps, to these unrealised sorrows of my childhood, but I hope it will be a harmless one.

The blacksmith's shop was an important ancillary to the boatbuilders' yard and I should imagine it was entirely given over to work on or for the boats. There was a typical 'village blacksmith's shop' in Anstruther and there we used to watch horses being shod and iron bands being fitted on to wooden cart wheels, but the blacksmith at the harbour was a dour man, who did not tolerate his own children in his workshop, far less their friends. But we had one contact with the blacksmith, strictly in the line of business. It was he who made our hoops or 'girds', and the very mention of them takes my mind clean away from fishing and industry to the real business of life for a child – games. These will be described in a later chapter.

Facing the harbour, just across the Harbour Head, was the sailmakers' loft, which was a very busy place when I was young. We children were not too welcome there either, but as the sailmaker's family were our schoolmates and very popular, we occasionally got into the loft, where the men sat cross-legged in front of the open shutters and stitched the greyish white canvas. Only where the men sat was there light and the shadowy corners of the loft with their piles of sails in various stages of manufacture or awaiting repair seemed mysterious and rather eerie to me. Behind the shop and loft was a large paved yard with boilers, where, I presume, the sails were 'barked' and stretched, but I don't remember ever seeing this process. The yard was for us a place where the gifted Johnston family staged 'shows'. If our seaside had no Pierrots or other professional performers, apart from the periodic visits of Punch and Judy and an occasional German band or dancing bear, we had the Johnston Shows. Everybody with any talent at all was given a solo part, and the rest of us, who could not lay claim to anything special, formed the chorus of the crowd and generally made ourselves useful. Our chief use was to sell tickets to all our aunts, cousins and other relatives. Mothers were no use; they wanted to know what we were up to. So we dressed up in all the finery the Johnston ragbag could produce, and we paraded through the town, the leaders carrying a hand-printed poster with the whole bill of fare, followed by a 'band' blowing penny whistles (they actually cost a penny!), banging saucepan lids together and drumming on biscuit tins.

Behind the band came the small fry with handfuls of handwritten tickets – 3d. adult, children 1d. The proceeds were of course for charity; but charity begins at home, and after all the artistes, stage-hands and hangers-on had been entertained to fizzy lemonade and biscuits, there was seldom any money left.

There was also a rope works, or 'rope walk', known locally as the 'Roperee', with the accent on the last syllable. My memories of this are so vague that I am not sure whether they are actual visual memories or just the pictures which came into my mind as Mother told us stories of her young days. But I know where the 'Roperee' stood and I see in my mind's eye the long low building stretching from the railway bridge to the railway station, separated from the lines by the station approach. Both ends of the building are open in my picture and the parallel lines of the cords being twisted into cables run together at the far end, so distant is it to the eye of a child. I don't know when the 'Roperee' ceased to function, but it must have been about the beginning of the century; by the time I went to the Waid Academy in 1907 and walked along the Station Approach every school day there was no trace of it.

In East Green, just near the harbour, there was a cork factory, where I suppose floats for the nets and 'fenders' for the boats were made, but this went up in flames before I was old enough to know anything about it. It was a favourite story in our family, and Mother used to tell us of the danger to the gasworks, just across the narrow street, and how a pipe was quickly run the whole length of the East Pier and the gas discharged from the holders into the sea. The danger averted by this action was not at all real to me; what I liked best about the story was the account of the succession of bangs as the bales of cork exploded in the heat.

A little further inland, at the Back Dykes, was the 'Iceworks'. All I ever saw of that was an iron door, always fast-locked, leading, I presume, into some refrigerated storeroom, through which the blocks of ice were unloaded from the lorries. I just remember catching glimpses of the crane chains with sharp metal pincers which grasped the blocks and swung them into the dark, cold interior. When I read of instruments of torture, I always saw these chains and pincers and imagined what it would feel like to be laid hold of in this way and swung into limbo.

The factories were sited in Cellardyke. There was a net factory where the work was very hard, as each knot in the net was made by the operator, a woman or girl, jumping from one treadle to another. And there were two oilskin factories, where the fishermen's oilskin garments were made, and also the small buoys used on the nets to mark where the lines were shot or crab or lobster pots placed; these latter were fixed on poles. The whole process of manufacture went on in these factories, from the oiling of the cotton to the processing of the finished garments. These factories changed with the times, and when the demand for fishermen's gear declined, they began to make rainwear for civilians and later, knitwear and other popular lines, so that they still continued to provide work for the women and girls. But the net factory ceased to operate before the fishing declined, probably because of the introduction of more efficient mechanical machinery in other net factories elsewhere which did away with the hard labour of the local factory.

One of my most vivid memories is of a disastrous fire which destroyed one of the oilskin factories. I was taken up to Kilrenny Road, from which a good view could be had of the close cluster of narrow streets which made up Cellardyke, and in the midst of these homes were the blazing factory buildings, with the flames shooting high into the air as the stores of oil were consumed. As a child I had a clear visual appreciation of everything I read or heard. After this incident, and whenever we had the Old Testament story of Abraham's interview with the Angel, when he interceded for the inhabitants of Sodom and Gomorrah, I knew where that interview took place. It was there on the Kilrenny Road, and what I saw when the factory burned was what Abraham foresaw and dreaded, the fire which consumed the sinners of the cities of the Plain. The fire-fighting apparatus of these far-off days consisted of a hose on a handcart, pulled by the members of the fire brigade with their brass helmets on their heads and hatchets in leather belts round their waists. They made quite an impressive show on ceremonial occasions, but were pitifully inadequate when faced with a fire of such dimensions.

There was also in Anstruther Easter an aerated water factory which had little connection with the fishing, unless it were to provide a substitute for the stronger waters supplied in the pubs. This was sited just behind Anstruther Town Hall and almost adjoining the fish-curing yard where my father was employed. Had it not been for this proximity I doubt whether I would have known of its existence, but on the occasions when I took my father's tea to the yard I was sometimes lucky enough to pass this factory when the hatches on to the road were open. I could then look down into the interior, several feet below the level of the road, and to my eyes almost in the bowels of the earth. I don't remember anything about the machinery or anything else in the factory, for all my attention was rivetted on the men with their wire masks, a precaution against exploding bottles. I knew all the men who worked there, and I went to school with their daughters, but the faceless figures moving in the shadows of their subterranean cave had no connection in my mind with the names I knew quite well. This small factory was known locally as the 'Breweree' from which I gather that it had at one time brewed beer or other alcoholic drink before it had been converted. Its situation was doubtless accounted for by the existence of a well just at the foot of the Birlie [burial?] Braes.

There was also a flour mill in Anstruther, situated just where the Dreel Burn became an estuary. I remember watching the water-wheel go round and playing, rather dangerously, on the banks of the dam, but I have no recollection of seeing the interior of the mill or any of the processes. It must have ceased to function as a mill about the beginning of the century, and the sheds and mill yard were converted into storage premises for fish-boxes and other fish-curing gear. The mill race became choked with weeds, the dam almost dried up, and the sluices and water-wheel slowly disintegrated. The 'Burns', as we called the sloping grassy banks between the burn and the mill race, which were among our favourite recreation grounds, became derelict, and even the well-worn paths vanished.

7. Economics

At the time of which I am writing I would not even have known the meaning of economics, so I cannot pretend to give a detailed or an accurate report on this aspect of the life of the fisher folk. I don't suppose I understood what 'half-dealsman' meant, except that it was a disparaging term when someone said: 'So-and-so is marrying a half-dealsman'; or what might be the significance of the appearance of truly magnificent pies – as big as a soup plate and costing the fantastic price of one shilling or one shilling and sixpence in the baker's shops at the end of the fishings. Nevetheless, such scraps of information fitted naturally into the pattern as the child's mind became capable of appreciating such things as wages, earnings and profits.

All the boats were privately owned, usually by a family, probably built for one man, and bequeathed by him to his sons. The crew consisted of six men and a cook, and only the cook was an employee; all the others worked for shares. Each of the crew owned one-sixth of the nets. This was in the days of sail. When steam drifters, and later motor boats, began to operate and eventually to supersede the sailing boats, the pattern broke up and to me, at least, was no longer clear-cut and distinctive.

Each fishing was treated as a separate venture and earnings were not distributed until the end of the fishing. Then the total earnings, presumably after expenses of running the boat, such as wages of the cook, harbour dues, etc., had been paid, were divided into two – one half for the boat and one half for the crew. This half was then divided into twelve shares, one for each member of the crew, and one for his share of the nets. The sixth part of the crew's earnings was called a deal, and it was quite customary for a man to be engaged as a member of the crew who had no nets of his own or capital to buy them. In that case the skipper, usually the owner of the boat, or some other member of the crew, would supply the necessary nets and draw half of the other man's deal in payment – hence the term 'half-dealsman'.

Book-keeping was non-existent. The fish-buyers acted as agents, collecting the proceeds of the catches and paying bills during the fishing, and at the end of the fishing the balance was paid over to the skipper in cash. And it was cash in those days – bags of sovereigns, half sovereigns, silver and copper, though banknotes were increasingly used instead of gold. Then all the crew met at the skipper's house for 'the pairtin' and after the man-sized pies had been consumed, washed down with strong sweet tea and doubtless with a dram in non-teetotal houses, the share-out took place. First, the cash was halved and the boat's share disposed of according to ownership. Then, like dealing a hand of cards, the notes or gold coins were dealt round in twelve piles, followed by silver, and last of all copper coins. Then and only then, would a man know how much he had earned in the past eight or ten weeks. But he was not allowed to gloat or gloom over it for very long, for the women were the financiers in fisher families.

As the men came home from the skipper's house, the wife and mother would sit at the fireside wearing a big white apron, and into that apron went the 'deal' of every man of the household, husband and unmarried son. If the woman was a good manager she would have anticipated the receipt of the money by getting in the bills and budgeting for the needs of the household until the end of the next fishing, and the balance, if any, would be paid forthwith into the bank. But not all women had the will or ability to manage money in this way, and some of the shopkeepers, who had financed such households on credit while the men were at sea, would be left to whistle, or even to take legal proceedings to recover the money. A neighbour of my mother, before she married, was overheard to tell a grocer who had called at the door to ask for payment of his overdue account: 'Ye're ower late; it's a' in the bank.'

It may seem strange that the unmarried sons handed over the whole of their earnings to their mothers, but it was in fact a very practical and sensible arrangement. The mother provided and maintained the nets, and when the son married his mother paid for all the furniture, except for bedding, bed- and table-linen, china, pictures, ornaments and rugs which the bride had been accumulating since childhood. As early marriages were the rule the son usually had the best of the bargain.

While at sea and when the nets were 'shot', the men whiled away the tedium of the night watches by 'jiggering' herring – that is, catching the fish on hooks as they surfaced. These herrings were usually of the best quality and fetched a higher price than the netted fish, and each man kept and sold his own catch separately. This was his pocket-money, and for this he was accountable to no man (or woman).

The fisher folk led gamblers' lives, not of choice but of necessity, and no doubt this explains the extremes of wealth and poverty in a small community where, as far as could be, there was equality of opportunity from the start. Of course there were 'lucky' boats and lucky skippers, but even a lucky boat could not long support womenfolk whose only idea of dealing with a lapful of wealth was to spend it all in the shortest possible time on 'braws' for self and children. The standard by which clothing was measured was how much it cost. As one woman said of her daughter's dress at a wedding or some such function: 'It may not be the brawest there, but I'm sure it is the dearest.' But at the other extreme there was the wife who could clothe and feed her household adequately even after a bad fishing, when a deal could be counted in units instead of tens or hundreds of pounds. This was the most marked difference between the fisher folk and the others. A fixed wage must have seemed a miserable pittance after a good fishing and an unattainable security after a bad one, but in any case it had no possible place in a fisherman's scheme of things.

There is no doubt that this simple plan of profit-sharing would have changed radically in the changing world of the first half of the twentieth century, but it would have been a gradual change, scarcely noticed from year to year, until a generation was born who knew nothing of the old ways of which its elders themselves had by then scarcely a memory.

There was in fact a depletion of the manpower of the fishing industry round about 1905–10, caused by a nation-wide emigration to Canada. I remember it chiefly because of the farewell meetings held in the Baptist schoolroom when we had tea and buns and tearfully sang: 'God be with you till we meet again'. Most of the young men I knew personally were tradesmen, and I do not know whether many young fishermen left the district at that time, though in recent years I have heard of settlements on the Great Lakes and in British Columbia where Scottish fishermen still form separate and distinctive communities and are still engaged in fishing.

But 4 August 1914 brought the changeover in a matter of days instead of years or decades. The boats were fishing in the North – Aberdeen, Peterhead and the Moray Firth – and it was a good fishing, plenty of herring, good prices and the prospect of several more weeks of good catches. I doubt whether the fishermen were even vaguely aware of the course of events on the Continent. I suppose it is difficult for this generation to realise how isolated small communities were in the days before the miracles of radio and television brought the world to our doorstep. These men were at sea or asleep in their boats for most of the twenty-four hours on six days of the week. There were no Sunday newspapers, even if there had been any among them so curious and irreligious as to read one. So suddenly, without warning, the boats were ordered to return to their home ports; the men in the RNVR or the Territorial Army received calling-up papers, the banks declared a moratorium, and the whole order of life was changed, never to be restored in the same pattern.

I was at that time an apprentice shorthand-typist with a firm of solicitors whose partners were also agents of the local branch of the National Bank of Scotland. This was only one bank of the three in Anstruther, and all of them were bankers for the three fishing communities of Cellardyke, Pittenweem and St Monance. Mr Watson of our firm had one of the few private motor-cars in the district – an imposing chauffeur-driven Daimler limousine – and after some frantic telephoning to the head office in Edinburgh, he set out to bring down specie to pay out the proceeds of the North fishing to the fish-buyers who banked with the National. In these days, when the weekly wage packets of one factory made a sum worthy of a wages-snatch, £60,000 seems a small amount to make a fuss about, but as it was unloaded in the closed bank office it looked to me like all the money in the world. After I had spent hours making up £100 bags of silver into twenty £5 bags (and almost inevitably being left with one coin too many or too few for the twentieth bag) I could well have opted for an economy based on barter and not on money.

The next scene in the drama, after the boats had come home and the money was all paid over, was the arrival of a gentleman from the Admiralty to commandeer the entire steam-drifter fleet (or as many as were seaworthy) as minesweepers. I had the good fortune to be lent to him as his secretary for the operation. He spent the first half hour or so indoctrinating me into the secret nature of my duties. He kept on repeating: 'You are a mere machine; you don't know what you are typing.' After I had fully realised the responsibilities of my position, he proceeded to dictate a series of lists, letters and schedules in no way different from my daily work, except that the details referred to steam-drifters instead of houses, farms, dispositions (*anglicé* conveyances) and bonds and dispositions in security (*anglicé* mortgages), and before being entrusted with these everyday matters I had had to sign a document binding me to absolute silence concerning anything which should come to my knowledge in the course of business activities. I trust that the passage of fifty-two years since I made these solemn promises has released me from my bonds.

I must have satisfied the requirements of this august gentleman, for after he had left to return to Whitehall he wrote an appreciative letter to Mr Watson in which he asked him to give me five shillings (more than a week's wages at that time) 'to buy a bit of blue ribbon to tie up her bonny brown hair.' Of course it may only have been my complete innocence (or ignorance) which kept our association within strictly business lines. If only I had seen a television programme showing how secretaries are expected to behave, especially on 'secret' assignments, I might have made more than five shillings out of the experience.

So the steam-drifters went, all except a few wood-built ones which were too old to be considered serviceable in the Navy. The sailing boats were not allowed beyond territorial waters and a grim and almost desperate situation faced the fishing communities. Most of the skippers and crews of the steam-drifters were recruited with their vessels, and for them the economic situation was not too bad, but for the older men and the unfit among the younger ones it was catastrophic. The unemployment of the 1930s could not have been imagined then, and it was an unheard-of thing for hundreds of men to be thrown out of work without notice. Some of the older men used the yawls and small boats and did some line fishing or 'went to the creels' (i.e., crab and lobster fishing) inshore, but in Cellardyke this type of fishing had always been considered as next door to the scrap heap, and often from inexperience, dissatisfaction, or genuine bad luck they made a very poor living. In course of time some of the younger men found employment at the naval base at Rosyth or elsewhere in jobs which had some connection with the sea and boats, but many more gave up the sea altogether. The drift from the fishing industry was to become an avalanche, and for a great many there was no return.

These desperate conditions did not continue for the whole of the First World War, but I was so much involved in the machinery of local government adapted haphazardly to matters of national importance – national registration, military service acts, with local tribunals, rationing of food and fuel, all completely novel and untried – that I had neither time nor inclination to concern myself with the general affairs of the fishing industry. Of particular details I had more than my full share, as the greater part of the population of Pittenweem, where I was installed in 1915 as clerk in the

Town Clerk's office, was engaged in fishing, and I had to deal with all the form-filling which these matters involved. I was also required to deal with the completed forms when I had extracted the necessary information, mostly from the womenfolk, as I have explained earlier.

The most surprising development, which presaged the end of the traditional partnership of boats, gear and men, was a strike which took place just after 1918, when some of the steam-drifters were released from government service. It was a very small affair as strikes go nowadays, and I doubt whether it was known outside Anstruther or Cellardyke, where it started, but in that small community its impact was terrific. I never knew the rights or wrongs, or even which interests were arrayed against which, but the strike certainly destroyed what faint hopes there might have been that fishing would be resumed as in the days before the War.

Although most impressive locally, the strike was not the only or even the most important reason for the decline of the fishing industry. The grim fact was that the continental market, the mainstay of the fish-curing trade, had gone completely, never to be restored. The greatline fishing for white fish had already been taken over by trawlers which could fish in Arctic waters and stay at sea for two or three weeks at a time. The day of the steam-drifter was over, though no one would acknowledge it in 1918. The Government made loans to the fishermen so that they could buy back their boats, and the fleet continued to operate in dwindling numbers until the 1930s. By this time the high running costs of steam-driven boats made it difficult to show a profit unless the fishing was good, and for some unknown reason the herring shoals appeared to change their habits and fishings were bad. Motor-boats took over from sail and steamboats, using seine nets instead of drift nets, and the whole pattern of the fishing changed. This new-look fishing community with its sale rings, fixed auctions and market facilities, is not within the scope of this book.

8. Customs and Superstitions

NOT ALL OF THE CUSTOMS of the time were unique to the fisher folk, but most if not all of them seem to have disappeared in the past two generations, so it may be of interest to mention them.

To begin where life begins, at the birth of a child. Every person calling at the home where there was a newborn child was offered the 'bairn's piece', and for this purpose sultana cake, scotch bun and shortbread were bought or baked before the birth. Liquid refreshments were also offered, no doubt a dram in non-temperance homes, but in the milieu I knew best fizzy lemonade was the only alternative to the usual cup of tea.

At the christening the first person to greet the mother and child after the ceremony was given the bairn's piece and a silver coin, the value of which varied with circumstances. As the christening always took place in the home, the recipient of the bairn's piece would be one of the family or friends, probably a child to whom the threepenny or sixpenny piece would be wealth. After the First World War it became fashionable to have the christening in church, and so far as I know the bairn's piece was discontinued. But I read in a newspaper some time between the wars of an incident in Dundee which a local correspondent thought sufficiently interesting to pass on to the national newspapers. He reported that one Sunday morning a visitor to the town happened to be passing a church when the congregation were leaving after the morning service. Out of the church came a proud young couple with their first child arrayed in a long christening robe, supported by grandparents on both sides. The visitor, coming face to face with the little procession and being interested in babies, stopped to admire the infant. She passed the usual compliments on the beauty of the child and the embroidery on the robe, and was about to move on when the mother pressed on her a bag of assorted cakes, while the father produced a half-crown which they insisted she should accept: to refuse the bairn's piece was to bring down ill-luck on the child.

Eatables took chief place too in the wedding customs. For a week or so before the wedding the bride's parents kept open house. Anyone was welcome to call and see the bride's 'providing' – piles of rugs, stacks of pictures, china and pottery of every description, and items of bed linen by the dozen. Some would bring a present or 'minding', but for the convenience of the casual passer-by or the neighbour who might not be able to afford much, there was a plate placed by the door, and offerings, large or small, could be placed there. For everyone there were provided the same refreshments as ushered the newborn child into society.

Weddings were usually held on Friday or Saturday, in the evening and in the bride's home, or in a public hall if more guests were invited than could be accommodated in the two-roomed house. To English readers it may seem a strange hour and place for the solemnization of a marriage, but in Scotland the ceremony is not hedged round with permitted hours and places licensed for the purpose, so 7 o'clock in the evening was the usual time, with supper at 8 o'clock. For the rest of the night, and, if it were a Friday night, well into the early hours of the morning, everyone, including bride and groom, would participate in dancing or some other form of entertainment. For several hours before the ceremony the bride in her wedding dress sat in 'the room' and all and sundry, neighbours and friends, called in to see her, to finger the stuff of the dress, discuss the merits of the dressmaker and compare style and price with other recent wedding dresses.

The supper was the really important part of the wedding. The menu was identical in every case – 'Dyker ane, Dyker a'' again – and consisted of beefsteak pies and fruit tarts which were baked in the baker's oven, whether prepared at home or by the baker. There would be the usual vegetables, custards, jellies and what not, and heaps of 'fancy bread' and bakers' cakes, but the pies, savoury and sweet, were the main and unalterable features. At the supper the bride sent out the 'bride's plate' to the oldest relative on either side – usually a bedridden granny or 'dey' [grand-dad]. A portion of everything served at the meal was heaped on to a large plate by the bride and carried by the bridesmaid to the proper recipient of the honour, and woe betide the unlucky bride who made a mistake in precedence.

The boat in which the bridegroom was a member of the crew was dressed with flags and bunting, and possibly the menfolk had their own equivalent of the 'stag party', but of this I have no knowledge.

On the Sunday following the wedding the whole bridal party – bride, groom, parents, bridesmaid and best man – sat together in church for the 'kirking', and that completed the celebrations. Honeymoons were unknown, or known only as something 'frem', i.e., foreign to fisher life. As likely as not on Monday the bridegroom went off to sea and the bride settled down to the job of running her own home and looking after her man's gear.

Funerals were held on Sunday afternoons, for only then could it be certain that the men were ashore, and only men attended funerals. But before the funeral there was the chestening, when the body was put in the coffin and the lid screwed down; as a matter of fact only this last act was performed at the chestening. The minister attended and conducted a short service, and the womenfolk were present then. This chestening service was not confined to the fishing community and so far as I was know it is still carried out in some parts.

Invitations to a funeral were by word of mouth and were delivered, I believe, by the beadle of the church. There was nothing ceremonial or solemn about such an invitation. The kitchen door would be pushed open, the beadle's head would appear, and as soon as he had uttered the words: 'Ye're bidden to Jock Tamson's funeral on Sawbath at half-past two', the head was withdrawn and he was on his way to the next house. When there might be upwards of 200 people to invite, there was no time to linger.

Mourning of course was *de rigueur* and strictly regulated. A widow was expected to wear a crêpe bonnet and veil, as well as a crêpe-trimmed black dress for the first year after her bereavement, and the stages by which such mourning was shed were just as strict. All the children had to have everything black, at least for Sabbath wear, though in the interests of economy dark-blue clothes could be worn to school, or old Sunday clothes dyed black.

I do not know whether the superstitions I encountered when I was young were confined to the fisher folk, but it was among them that I learned of these things, and I cannot claim to have any exhaustive knowledge of the subject.

Some of the things which were regarded as unlucky were undoubtedly connected with poaching. It was unlucky to mention salmon; such fish were known as 'red fish'. Partridges were always referred to as 'broon do'es' [brown pigeons], but I don't know why it was forbidden to name a pig. That animal was always described as 'cauld iron'. There was an amusing story current when I was a child concerning a local character of an earlier generation. This man had killed a pig belonging to Innergellie farm, but the laird was unwilling to prosecute him and offered to let him off provided he called at the 'big house' and apologised. A fine or a few days in jail would have terrified Jamie less, but having been carefully coached by his friends, he presented himself at the back door of Innergellie House, prepared to recite his piece. When the laird himself appeared, Jamie's wits and prepared words deserted him, and, crimson faced, he blurted out: 'Beg my pardon Mr Hendergellie, it was me that killed your cauld iron.'

It was very unlucky to meet an old woman or a minister of the gospel on the pier or on any approach to the harbour when the men were on the way to sea. Few old women were in the habit of strolling out of doors at any time, and none of the fisher women would do such a thing. But there was one old crone who delighted in walking to the end of the pier just as the tide turned, so that she would meet some of the crews as they made for the boats. She was one of those women referred to in an earlier chapter who liked a dram and did not mind going, unobserved if possible, to the public house. Sometimes the men turned back, made a quick turn round by a street behind the harbour, and reached the pier after she had left. If necessary they would miss a tide rather than go aboard straight from a direct encounter with her. Her behaviour was just spite, taking out on these men the unhappiness of having lost her own man many years previously.

Another instance of bringing ill-luck on a fishing arose out of ignorance and a desire to do the right thing. When the boats left for the 'South fishing' in September it really was an occasion. It was then that all the children and old men thronged the piers and the fishermen handed out boat's biscuits for luck. A young minister had just come to town, and seeing the crowds and wishing to take his place in the life of the community, he strolled down the pier, calling out greetings to all and sundry and hailing members of his congregation by name. He was answered by dead silence, and one by one the crews came off the boats and made their way home. The minister was not on the pier when they left next day, nor ever again, if boats were putting out to sea. To welcome them home and get a fry of fish or herring was quite a different story.

There were, of course, all the usual superstitions – things that were lucky or unlucky – but in our household at least no one paid much attention to them. Only one sticks in my mind: green was the unlucky colour – 'After green comes grief' – and it just so happened that at the time of my father's death, when I was nine years old, my new Sunday coat was green, and it had to be dyed black.

But there was one magical rite in which I thoroughly believed and practised on more than one occasion. Among my class-mates was a girl from a tinker family who abandoned their covered cart in the winter months and rented a house, so that the children could attend school. Mary Cassidy had all the charm of the gipsies and we were all fascinated by her stories. She was not only a tinker, or 'tinkie', but she was a Roman Catholic, and so we were really risking hell-fire to listen to her. The magic rite into which she initiated us was a means of finding, or rather making, gold; not the transmutation of metals, but changing bones into gold. The bones had to be human bones, gathered from the beach when the new moon first appeared. They were buried in the sand below high tide-mark and the incantation was pronounced by Mary. I don't think any of us learned the words; I know I did not. The bones could not be disturbed until the next new moon, when we carefully searched the sands. Needless to say, we found no gold, nor even the bones we had buried, and Mary told us that the reason was that they were not human bones, but only the bones of some animal. She may have been right; our beach was not usually strewn with human skeletons, but I doubt if even human bones would have remained in the same place after being washed over by fifty-six tides!

Hallowe'en customs were the same among the fisher folk as elsewhere in rural Scotland. For several days before 31 October bands of children dressed in finery from the household rag-bags, with blackened faces or papier mâché masks [fausse faces], the boys carrying 'neep' [turnip] lanterns, made a round of the homes of friends and neighbours. They were known as 'guysers', but had no connection, so far as I know, with Guy Fawkes and 5 November. Celebrations of Guy Fawkes' Day were completely unknown to us. The 'guysers' when admitted were supposed to give a performance of some sort, and probably there may have been something organised among the older set in connection with the Hallowe'en parties, but I only remember 'guysers' when I was a small child, and they were all small children who giggled and stammered, solo or in unison, as they rendered some 'poetry' learned at school or sang some school song. Having 'done their turn' the children would collect their pence and proceed to the next house, where the performance would be repeated. There were no carol singers, but the 'guysers' bore a slight resemblance to them.

The Hallowe'en parties were great fun, and here the fisher bairns had the best of it, for such parties were very messy affairs and the garret was the only possible place to hold them. First of all there was ducking for nuts and apples. Only very small children were allowed to hold a fork or spoon between the teeth while hands were clasped behind the back; all others had to pursue the prize with open mouth to the bottom of a deep tub of water. Very soon the floor was awash and all the participants in the game were completely soaked. But the open fireplace was piled high with glowing coals and one soon got dry while eating the hard-won fruit. At this stage the game with nuts was played. Lads and lasses were represented by nuts, placed in pairs on the 'ribs' or fire-bars, and according to the way they behaved under fire, their future life together would be foreseen. Very few couples caught fire and burned quietly side by side; in most cases one exploded in a shower of sparks and the unfortunate lad or lass was sure of a stormy married life.

By the time this game was over and everybody warm and dry, the 'chappit tawties' [mashed potatoes], the *pièce de résistance*, was ready. This was prepared by the adults while the children were 'dookin' and eating their apples and nuts, and the big black pot was placed in the middle of the floor. Everybody sat around, and the lights were put out. In the light of the fire only the meal was eaten from the pot with fingers – nature's forks. All sorts of surprises found their way into the pot, some pleasant, and some not quite so nice; but all was eaten with gusts of laughter. This laughter reached a climax when the lights went on, for in the dark the lads had been stroking the lasses' cheeks, having first rubbed their hands on the sooty sides of the pot. How we rolled over the floor in ecstasies of mirth as we saw each other one by one – and ourselves reflected in a mirror!

Then scones, some spread with black treacle and

others dipped in flour, were hung from the rafters. These had to be bitten as they dangled, while the hands were clasped behind the back. Again the cosmetic effect of these efforts raised shrieks of laughter.

By now the 'clack' [treacle toffee] was ready to pull, and many strange shapes were evolved at the expense of blistered fingers.

At this stage the children were packed off to bed and the adults took over, but by the time I reached that exalted state my interest had waned and I had ceased to be invited to Hallowe'en parties, so I was never initiated into the mysteries.

As I said, we had no carol-singing in my childhood; Christmas Day was not even a public holiday and I remember quite well comparing notes with other children at morning school as to what we had found in our stockings. We were usually given an *ex gratia* half holiday, and the New Year's holiday – a week free from school – began next day.

New Year's Day was the real holiday of the year, and for the children it began with first light on 31 December. That was Cake Day, when every house was open to children, and even the National Bank had 240 newly-minted pennies, as bright as gold, ready to give to the first 240 children to call there. This was through the generosity of Mr Watson, the agent, and I never knew or enquired whether other banks did the same. The bakers made special cakes or biscuits known as 'pentecuts', some of shortbread with the selected child's name in icing sugar, some of a plainer biscuit type, but all of the same shape – a quarter of a circle. In each of the corners a small white sweet was embedded, which was the seal of the genuine pentecut. Grocers stocked up with cheap oranges and rosy apples and every house laid in a stock in readiness.

The traditional words to be said at each call were:

Rise up auld wife and dinna be sweir
But pairt yer cakes as long's yer here.
The day will come when ye'll be dead
Ye'll neither care for cakes or bread.

or

Ma feet's cauld
Ma shure thin
Gie's ma cakes
An' let me rin.

In practice it was more usual to say: 'Gie's ma cakes.' The bairns who had a large circle of friends carried pillow-cases – not the white flimsy ones in use at home, but the strong, dark-blue and white striped ones belonging to the boats. By the middle of the forenoon they would be struggling home with the spoil. My mother, worthy descendant of Ann Fleming, thought it was a form of begging, and, as we were really poor, she would never allow us to join in the fun. Only folk who do not need to beg can do so and still keep their pride.

The forenoon for feasting; the afternoon for fireworks. These were not the elaborate, costly variety of the present day, but just halfpenny or penny squibs and bangers. No doubt these noisy objects had the same original purpose as that which led to the discovery of gunpowder by the Chinese – to scare away evil spirits – but the immediate effect was to scare all the dogs in the vicinity and to compel timid womenfolk to keep indoors.

Of course it was a day that did not end, for on that night of the year no one was ordered to bed. A child might not be able to prop his eyelids open and would be laid down while the fun went on all around him, to waken on a New Year. Surely there never was a day like New Year's Day! As a matter of fact I remember very little of what did happen on New Year's Day, and I fancy that most of the bairns were like me, too tired and exhausted by Cake Day to bother much about New Year's Day. It was only for grown-ups anyway!

Easter we celebrated by rolling eggs down the braes – hard-boiled eggs, dyed or painted with bright colours. When they broke we ate them and that was usually the first picnic of the year, and a very cold one usually. This celebration did not necessarily coincide with Easter Monday. In fact in our district it only did so when Easter fell right at the beginning of April, for the first Monday in April was our spring holiday and neither Good Friday nor Easter Monday was a public holiday, nor are they to this day in Scotland.

Whitsun was just one of the two 'flitting' terms, days on which rent was due and tenancies began or came to an end; the other was Martinmas. Whitsun was a fixed date, 15 May for payment of rents, 28 May for 'flitting' or removing.

Here I think we might have a look at the games calendar, as fixed as any church calendar. The timing of bird migration is not more wonderful or mysterious than the impulse which produced the proper equipment for the seasonal games spontaneously and simultaneously. 'Girds' [hoops], tops, 'cat and bat', kites, balls, skipping ropes, 'bools' [marbles] and buttons, and 'chuckies', all appeared, each at their own appointed time: of who or what fixed these times I have no knowledge.

All our toys were of the simplest and cheapest kind, and mostly home-made; fortunately for us the prestige value of toys had not been discovered. There were of course a few shops which sold toys, dressed dolls and the like, but these were not for everyday use. 'Girds' [hoops] as I have said were made by the blacksmith. These were essentially a rounded ring of metal, trundled by means of a stick, if you were an inexperienced boy or a 'mere girl'. The real experts took their girds back to the blacksmith and had an iron rod attached to the gird by means of a small ring, so that the rod could be held steady at the correct angle in order to bring a constant pressure to bear on the spinning hoop. These were real record-breaking girds, both for speed and distance. I don't think I ever achieved this skill, though I 'borrowed' my brother's gird whenever opportunity offered and tried hard to learn. Girls, in our family at least, had no girds of their own.

Tops were made of wood, machine turned I suppose, and retailed at a halfpenny each in stationers' shops. Before they could be used they had to be 'shod' with an iron tack which the shoemaker would knock in for nothing. Then the rounded surface of the top had to be decorated in colours so arranged as to make a definite pattern when the top was spinning. Coloured chalks were most useful for this purpose, because the pattern could be altered at will. To propel the top there was a 'last', a length of cane with a piece of thick string or rope attached to one end. The string or rope was frayed at the end to make a proper 'tail' and the manufacture of these lashes, which had to be the proper weight for the particular top, was quite an art, and one in which my mother excelled, as she also did in making and embroidering balls, and making well-balanced kites which would fly straight up and head on to the wind without tumbling about or crashing. How my mother found time to make our toys and teach us how to use them is one of the mysteries I shall never solve, except in so far as I realise that selfless love has reserves of time and energy which the rest of us use on ourselves

The balls she made for us were made of used wool, wound tightly and regularly so as to make perfect spheres, but with sufficient resilience to make them good 'stotters'. When the ball was of the correct size it was finished off by threading the end of the wool through the ball by means of a darning needle, and then the outside was embroidered in bands of brightly coloured wool. By the time I was promoted from the infant school to the 'big school' (at about the age of seven I suppose) india-rubber balls began to make an appearance alongside the tops in the stationers' shops, but at twopence each they were a bit beyond our means. In any case they did not last nearly so long as 'mother's make'.

Kites were for boys only, though a favoured sister might be allowed to hold one when it had reached full height. There was an obvious and rational season for kite flying, when the harvest had been 'led' and the stubble fields gave ample scope for manoeuvring. They were made of thin splines of wood tied together in the shape of a cross, the traditional cross-form with the crosspiece about one-third from the top. From end to end of the crosspiece a flexible piece of wood – usually part of a barrel hoop – was curved round and joined to the upright at the top. How often did I see Mother balancing this framework on one finger, paring a little here or there, until it kept its position. Then strings were tied to the ends of the half-hoop and fastened to the bottom of the cross. This frame was covered with newspaper on both sides, stuck together with flour paste. Then the tail, a long length of string with pieces of folded newspaper spaced regularly, was added to give stability. A very few box kites from shops made their appearance in the early years of the century, but no real boy would even look at them.

Skipping ropes were easily obtained where ropes were in constant use. Worn-out clothes-lines were the best, but the thinner brown cord used in the fishyards could be used for lack of something better. Sometimes a well-meaning relative would present a child with a 'shop' skipping rope with painted wooden handles and bells, but they were not long enough for anything but single skipping, and we mostly skipped in teams.

Dolls did not appeal to me personally, but for those who liked dolls – and there were many – there were wooden ones in various sizes, ranging in price from one penny to sixpence. They had brightly painted faces and black painted hair, and their limbs were jointed. These were dressed by their owners in scraps from the rag-bag, and were quite as precious as the wax beauties with hair and eyes that closed, dressed in the latest fashion, which stood, each in her separate box, on a shelf in the toy shop.

'Bools' [marbles] could be bought at one penny a

dozen in sweet shops, and big glass 'bullies' for one penny each, but most boys seem to have had an inherited capital stock which they carried about in a small cotton bag. When the boys were playing 'bools' the girls played 'buttony', and the ammunition for this game was abstracted from the button-bag, which had a place in every household; no garment was ever finally discarded without having every button removed and added to the stock. At the end of the season the winnings were returned to the button-bag.

'Chuckies' were easily obtained – five evenly-sized stones from the beach – and each girl had her set, which she believed was better than anybody else's. I remember seeing sets of chuckies sold in a box, cubes of brightly coloured earthenware, but I could never imagine anyone getting through the first simple throws with such angular and unwieldy implements.

'Cat and bat' were also home-made. The bat was just a straight piece of wood cut into a bat shape to suit the size of the owner, and the 'cattie' was a four-sided piece of wood, about one inch square and four inches long, sharpened at both ends. This was a team game, played by both boys and girls, separately, with batters and throwers, and rules which were strictly observed. It was rather like rounders, and that game was often played with the same bat and one of our woollen balls.

9. National Events, Politics and Elections

THE FIRST EVENT OF NATIONAL importance in my lifetime was, I suppose, the Diamond Jubilee of Queen Victoria, but of the celebrations of this event I have no personal recollections. I know, from hearing the story told a great many times, that there was an open-air function, probably in the Waid Academy Park, with sports, processions, bands, etc., and free refreshments for all children. My next older brother was the youngest member of our family present, and he was arrayed in a velvet suit. The refreshments included rhubarb tarts, and the small boy attacked his treat with such vigour that it broke in half and the sticky juice cascaded down the front of his new suit, which was never the same again. That was all I ever heard of the Queen's Diamond Jubilee.

The Boer War made little impact on the life of our community, but one of my school friends had a brother in the regular army who served in South Africa at the time. Every day at school we listened to stories of the wonderful exploits of Willie McLeod, and if only half of these tales were true, he must have contributed considerably to the ultimate victory. Though few were personally involved, feelings ran high and anyone suspected of pro-Boer sympathies was the victim of persecution. One of the pupil teachers in the infant school was the daughter of a local publican who was, or was said to be, on Kruger's side, and all of us, infants though we were, made it our business to show her how patriotic we were.

I remember vaguely hearing of the relief of Ladysmith and Mafeking. Who or what Lady Smith was and what she was relieved from I never knew or questioned. One of my most vivid memories of these very early years was watching the torchlight procession at the end of the war. It was long past my bedtime and I was too young to know the joy of anticipation, so the minutes seemed like hours as we waited in the dark. Then the magic thing happened. Along the Shore Street came the procession with all the smoky torches and bright lanterns reflected in the harbour. I have no idea what the procession consisted of, but I can still see clearly the effigy of Kruger, bearing a strong resemblance to his local supporter, which was destined to be burned in the bonfire which marked the end of the show. I, of course, was fast asleep in bed long before the climax came, but I have no doubt the celebrations were as noisy and boisterous as anywhere else in Britain.

The Boer War and its finances touched me in a strange way. There was a tax on tea, imposed as a war tax, and it took the form of adhesive stamps affixed to each packet – a sort of direct levy, I suppose. One day after the end of the war, my mother sent me to the grocer's for half a pound of tea, and she gave me the exact money, including the penny which represented the tax. The grocer gave me the tea, took the money and then with a smile gave me back a penny and said: 'Tell your mother the war is over now.' To my mind there was just no connection at all between his action and his words. He had given me a penny and that was enough to keep my mind occupied for a while. What should I do with such unexpected wealth? A halfpenny [bawbee] was the limit of our spending money and that usually had to be shared with a younger sister. Should I take the precious penny home and have it doled out in halfpennies and shared, or should I blue the lot in an orgy to be enjoyed all on my own? A penny could buy a whole bar of cream chocolate or four ounces of mixed sweets and all sorts of variations in between. What I ultimately bought I do not remember, but I know it was bought and consumed in strict privacy; I then took my belated way home with the half-pound of tea. I did not need to give Mother the grocer's message; she took one look at the package and then demanded: 'Where's the penny?' there was neither time nor opportunity to explain; when Mother was angry the only thing to do was to escape the smacks which punctuated the exasperated questions, so I went into the garden and repented bitterly the feast which had tasted so sweet.

I suppose the next event I should have noted was the death of the old Queen, but of this I have not the slightest recollection. I think perhaps that small children retain happy memories more than sad ones, and that may be why I recall the accession of Edward VII as an isolated event. The coronation was to be celebrated locally in the traditional way, with a great gathering in the Waid Park, and each child was to receive a box in the form of a medallion with the new King's head on it, and a chocolate medallion inside. They were gilt, and to us were real gold, so that I longed to posssess mine and dreamed of all I could do with it. Then the tragedy of the postponement of the coronation brought an end to all our hopes. The illness of the King and the anxiety of the nation meant nothing to us. The great day came and went, and weeks after, when the coronation actually took place I suppose, the gilt boxes were handed out at the close of an ordinary school day, and that was the end of that.

When Edward VII died I was in my last year at school, and I remember the service held in the Parish Church, which we attended as a school. There I heard for the first time the Dead March in *Saul*, and the cold shivers ran down my spine. By the time King George V was crowned I had started my career as a shorthand-typist and the occasion was marked by Mr Watson presenting each member of the staff with a newly-minted coin bearing the new King's head. This was meant to be kept as a memento, and for quite a while I did keep it safely. But coins were very scarce in those days, and there were so many things to buy that eventually it went the way of all the rest.

The sinking of the *Titanic* was the next event which stands out in my memory. There were no news bulletins to keep one up to the minute with what was happening in the world at large, and in our household we took only an evening newspaper. So I was quite unprepared to go into the office and find Mr Watson, usually the busiest of men, sitting at his desk gazing at *The Scotsman* with tears running down his cheeks. That shocked me more than the story of the disaster, which seemed so remote as to be unreal. Again the Dead March in *Saul* made me shiver, as in every church there was held a solemn service to give expression to the nation's grief. And yet within a very few years such sinkings became almost routine.

On the fateful first weekend of August 1914 I was in Edinburgh, and so I was more aware of the importance of those days that I would have been in East Fife. Sunday, 2 August, was the dreariest day one could imagine, and never surely did 'Auld Reekie' look so grey and grim. I remember nothing of the forenoon, so I may assume I went to a church service; any other programme on a Sunday morning would have been memorable. So it was only when I went out after dinner to visit a friend in another district that I became aware that something exceptional was happening. Newsboys were crying the headlines of a special edition – a very special edition published in the afternoon of a Sabbath Day! How can I convey the excitement and horror of such a world-shaking event? Indeed our world was being shaken and would never recover from the shock.

Nevertheless, though the world shakes, life must go on, so on Monday I carried through my pre-arranged programme, which was to buy a new outfit with my hard-earned pounds. I had found out beforehand where the best bargains were to be had, and so I did not waste time window-shopping; that was a pastime one indulged in with empty pockets. I made my way straight to a shop which had advertised in Saturday's *Evening Dispatch*:

'Ladies' Costumes in Good-Quality Serge – Twenty-five Shillings'. The year 1914 was one of strong primary colours in the fashion world: a subconscious preparation for the years of khaki, grey or black which lay ahead? The colour range in the twenty-five-shilling costumes was emerald green, royal blue, purple and tangerine. I chose the purple – a colour I have hated ever since – and to wear with it a hat of white straw trimmed with three tangerine roses (seven shillings and sixpence) and a pair of tan leather gloves (two shillings and sixpence), and went home well pleased with my bargains.

I had almost forgotten the shock of Sunday afternoon in the excitement of my shopping expedition, but that Monday was truly the last day of my old life. The moratorium which upset business life, the departure of the local Territorials (Black Watch, 42nd Highlanders) marching to the station behind the pipe band, the less spectacular departure of RNVS men, news that the fishing fleet was ordered home, the arrival of a company of soldiers to mount guard on the harbour and all its approaches, followed in quick succession. Not to be outdone, the local organisations took a hand in the defence of the homeland; Boy Scouts, with batons at the ready, relieved the guards on such important places as the railway bridge, and the local VAD [Voluntary Aid Detachment] organised an emergency hospital in the Erskine Hall to receive the casualties from a big naval battle which had already been fought in the North Sea. My superior in the office was a member of the VAD and therefore had to be given leave of absence. As our firm were local agents for the county families, whose ladies were officers of the VAD, I was commandeered to attend at the Erskine Hall to take an inventory of all the beds, mattresses, blankets, pillows and so on, which were lent by patriotic housewives of the town. So for me the war began as it was to continue, as a matter of work, work and still more work, which left me no time to learn about what was happening outside our immediate neighbourhood.

There was plenty happening there. Not to begin with of course. The great naval battle from which the casualties were to be brought to such a likely strategic spot as an improvised hospital in an Anstruther hall was of the same genus as the legendary host of Russian soldiers with snow still on their boots which later haunted every railway siding in the United Kingdom. The sentries who paced the end of the piers and challenged all comers had only to deal with flustered natives such as one who responded to the question: 'Who goes there?' with a stammered: 'David West going east', giving his real name and the direction in which he was proceeding, for all the fisher folk used 'east' and 'west' as the ordinary way of giving directions; only an idiot or a stranger (synonyms?) would not know that 'east' meant in the direction of Cellardyke.

But we lived at the mouth of the Firth of Forth, the entrance to the strategic centre of Scotland and the route to the Forth Bridge and Rosyth naval base beyond. All merchant shipping passed our doors, as it were, sometimes on the north side of May Island and sometimes on the south side. When the submarine menace grew we began to have grim evidence of warfare. My youngest brother, who had been one of the Scout volunteer sentries in August 1914, was by this time apprenticed to a firm of marine engineers whose works were situated, most improbably, in a country village three miles from Anstruther harbour. That meant a three-mile cycle ride in all weathers to start work at 7 o'clock, and boy-like he was always glad of any excuse not to turn out. It matters little what kept him from work one winter morning in 1915–16; whatever was his excuse he felt fit to take a stroll along the beach at the Billowness by the forenoon. When he came back to the house an hour later he looked more sick than he had ever been in his short life, for he had almost stumbled over the bodies of two dead seamen washed up on the rocks. They were the first of many, some stripped naked by blast, some having reached shore alive before perishing on a lonely stretch, trying to scrawl a last message on the rocks. We began to recognise and dread the sounds of torpedoes and to welcome the thuds of depth charges, and once I saw two tramp steamers blown up almost simultaneously.

The beaches were strewn with wreckage, sometimes with the cargoes of the boats. Once it was a cargo of oranges and the local doctors had to work overtime, dealing with children who had gorged themselves on such unlooked-for treasure. Occasionally, very occasionally, I took an hour off to play a round of golf on the nine-hole course which bordered the sea at the Billowness. One day I had played even worse than usual, which is saying something, and as I came in and banged my bag of clubs behind the door I exclaimed: 'I canna play golf for monkey nuts!' and Mother responded: 'Eh, lassie, is there another cargo washed ashore?' And then we both exploded in laughter as we pictured me literally ploughing my way through ankle-deep monkey nuts to reach the green. Thank God we could still find something to laugh about even if the humour was somewhat grim.

The armistice took us all by surprise. In Cellardyke the news got through to the factories about 2 o'clock, and all the girls marched out, as they did everywhere else, and began to celebrate in the streets. But I was working in Pittenweem, and there we had no factories to stage demonstrations, and so we each and all went about our respective duties until closing time. Only when I got home and found a married sister who had come along from Leven to rejoice with us did I learn that the war was over.

The next event, following very close on the armistice, was the general election of 14 December 1918. In order to put that into proper focus as far as the fisher folk were concerned I must give some outline of politics in the St Andrews Burghs Constituency of which Kilrenny [Cellardyke], Anstruther Easter and Anstruther Wester were important members, with a political importance beyond their numerical strength on the voters' rolls.

I was first aware of politics in a by-election in September 1903. The member whose resignation caused the by-election was the Hon. H.T. Anstruther, a brother of Sir Ralph Anstruther of Balcaskie, Lord-Lieutenant of the County. He had been elected in 1900 but I have no recollection of that election. I only remember Mother telling us about his small son and daughter being driven in a governess car through the streets, waving Union Jacks and calling out: 'Vote for Daddy!' In 1903 I was eight years of age and therefore old enough to sense the excitement that pervaded every department of life, even school and the playground. A cousin of mine, just two months younger and in the same class at school, came to me one day for guidance. She had just been asked if she was Tory or Liberal, and, as even the terms meant nothing to her, she came to me with the same question. 'Oh, I'm a Tory' I replied at once, and Mabel said: 'That's all right then. I must be a Tory too, for I'm your cousin.' As a matter of fact she was completely wrong. Her father, my mother's brother, was a prominent business man in Anstruther, a member of Chalmers Memorial UF Church, and therefore a Liberal. So when the election day drew nearer and we all wore our colours, Mabel had a rosette of red and yellow – a horrible clash we Tories thought! – while I and the rest of my family wore our patriotic red, white and blue.

So I discovered that politics did not depend on family relationships. What did they depend on then? So far as I could judge, the church you attended had a very great deal to do with it. Ministers of religion made no secret of their political allegiance; they rather flaunted it. During election campaigns political meetings were held every weeknight in one or other of the halls in the town and the chairmen of these gatherings were very often the local ministers. Our family worshipped in Cellardyke Parish Church and the Revd James Ray not only chaired every Unionist meeting in Cellardyke, but he also preached two Unionist sermons every Sunday of the campaign. So his congregation were Tories almost to a man.

Chalmers Memorial UF Church, on the other hand, may have been said to owe its existence to the Liberal party, for it was a former Member of Parliament –

Stephen Williamson (Liberal Member for St Andrews Burghs 1880–1885) who built their church, manse and beadle's house, thereby making them the wealthiest Free Church congregation in the district. To this church practically all the tradespeople of the two Anstruthers and the factory owners of Cellardyke belonged, and the majority were Liberals.

The Revd Peter Buchan, minister of the Baptist Church, was well to the left of the Liberals and a real fighter. What the Revd James Ray did for the Tories, Peter Buchan did for the Liberals, and on rare occasions, when they found themselves on the same platform for some cause which was supposed to transcend political strife, they found it very hard to observe anything more genial than an armed truce. So naturally most of the Baptists were Liberals.

Of course there were other factors involved, but a child of eight was hardly capable of sensing them. All we children knew was that 'the Major' (Major Anstruther-Thomson, afterwards Major Anstruther-Gray) was a real superman, huge in stature, bulk and reputation, whereas Captain Ellis was a mere shadow, lower in rank and in every other way; I don't know if I ever saw him, but I cannot recall a single descriptive item of this man, who in the event, defeated our hero and won the seat for the Liberals. The Major, on the other hand, seemed to be omnipresent. He was the topic of conversation everywhere, and we all knew of his exploits in the Boer War, where, according to Tory sources, he received a kick in the stomach from his horse when it was shot under him, which caused his obesity. The Liberal version was that his 'big belly' was the result of excessive drinking, and as they were, on the whole, on the side of temperance, this gave them a real weapon. We had our campaign songs of course, but only a fragment of one, complete with tune, remains with me. It went as follows:

> Vote, vote for the Major
> He's the man you want to know
> With his umberella hat and his belly full of fat
> He's . . . .

Presumably the last line told something of his or his party's policy or programme and that may be why it made no lasting impact on my mind. No doubt Captain Ellis's supporters also had their theme songs, but whether our side had louder voices in the playground or were more in number I don't know, 'Vote, vote for the Major' seems to have drowned all other voices.

The 1906 election has a clearer focus in my mind. I was now 10½ years of age and had serious views on many matters, including politics. I studied the huge posters which were on every hoarding and on the gable-ends of many buildings, there was the 'Big Loaf' and the 'Little Loaf' – Free Trade and Tariff Reform, and to me these were more than symbols; they were facts. Advertising was in its infancy, and minds had not yet built up the resistance which is needed for mere existence today, so we gazed at the pictures and faced the facts: Free Trade (Liberals) meant plenty for all; Tariff Reform (Tory) meant hardship and poverty; but when were idealists afraid to face poverty? Let the Liberals feast on their 'Big Loaf' and grow fat on it; we would tighten our belts and fight for our rights to 'keep oor ain fish-guts for oor ain clow-maws' [seagulls]!

This time the Major won (by only twenty-three votes) and his victory was marked by all the usual extravagance of the time. On election night the houses of prominent Liberal fishermen were coated with tar; windows were broken, and even more heads, as fights broke out all along the streets when the supporters went from the town hall to their homes. The Major arrived the next day in a 'carriage and pair' and a band of hefty fishermen met him at the railway bridge, took the horses from the shafts and pulled the carriage all through the two miles or so of streets to the Harbour Head at Cellardyke, where he was able to address the wildly cheering mob. What did it matter if the Liberals had won a landslide victory at Westminster? We had shown the world what we thought of them and their 'Big Loaf'.

But alas, in January 1910, when the Unionists almost made a comeback in the country at large, we suffered a defeat, and our Burghs was the only seat lost in Scotland. This was in spite of the fact that the Major was still the hero and his opponent only a barrister who was popularly believed, by the Tories at least, to have changed sides when he married a rich wife! By this time I was a pupil at the Waid Academy and very politically conscious. No hoarding-gazing for me this time – that was for kids. My brother and I attended every meeting we could, and our homework was sadly neglected, but as all the teachers were equally involved it did not matter much. I don't remember the issues very clearly, for our part was not to listen to the speeches but to cheer every Unionist speaker and to boo at every Liberal meeting. I believe that 'Home Rule for Ireland' and 'Support for Ulster' were the opposing slogans, and I remember one meeting addressed by the young, handsome Marquis of Breadalbane, when a row of teenage girls, including myself, captured the front seats of the gallery in Anstruther Town Hall and screamed our rapture in a way which would have put us in the popstar-fan class of the present day. How much damage we did to the cause we meant to support I'll never know, but I realise now that this large meeting of nearly 1,000 was the climax of the Unionist campaign, and our unruly behaviour probably ruined it. Perhaps we were responsible for our hero's humiliating defeat by thirty-eight votes.

Mr J. Duncan-Miller's reception in Cellardyke, when he made his tour of the constituency, was in marked contrast to the Major's triumphal progress in 1906. He arrived in a motor-car and was supported by members of his local committee from Anstruther Wester, of all places! Liberal 'Dykers' might have been tolerated; after all, one had to live with them now the election was over. But West Anstruther, the home of all the would-be gentry, who despised and ignored the fisher folk! It was asking too much to see their Provost basking in the sunshine of the immaculate barrister's fixed smile. The time too was ill-chosen – the forenoon, and a Friday forenoon at that, when every housewife was cleaning up for the weekend. News of the arrival spread along the streets as soon as the car had crossed Caddies Burn, and every doorway was filled by one or more women, arms akimbo, hands laden with whatever cleaning utensil they had chanced to pick up. It was a bitterly cold day with snow on the pavements and the car was open, so that all could see the new member. Who threw the first missile was never known, or at least never revealed. Possibly it was just a mischievous boy, for it was only snowballs to begin with. One hit the Provost full in the face, and the crowd, now practically filling the street and slowing down the car to walking pace, jeered good-naturedly: 'That's the first drink of water you've had in many a long day, Provost!' Had the Provost answered in the same vein all might have gone well, but he was a choleric little man, who saw in this taunt a deadly insult, and he made the mistake of showing his anger. This changed the mood of the crowd, and instead of snowballs blacklead brushes, pails, doormats and anything that came to hand hurtled into the car, which took the first turn available out of the main street and made its escape into back streets where the member and his committee could be tidied up and made presentable for the visits to the other Burghs.

Within a year there was another general election, leaving the position of the main parties practically unchanged, and the Labour Party now holding the balance with forty-two seats. No one could have foreseen that this precarious Government would last for eight years, almost to the day, with all the realignments made expedient by the First World War. For St Andrews Burghs it would be the last election, and, true to tradition, the pendulum swung right once more and the Major was returned with a majority of forty-eight, the biggest since 1900.

In the 1918 election St Andrews Burghs were merged with East Fife, which Mr Asquith had represented for all the years I could remember. I don't think anyone, not even Sir Alexander Sprot himself, believed that he could win the seat for the Conservatives from such a distinguished and long-established Liberal statesman.

There is no doubt that the unpredictable St Andrews Burghs' vote played a major part in this débâcle. We did not have our Major, but we had a soldier straight from the Field, where he had fought for four years alongside our boys, and he 'belonged' to the East Neuk with his estate at Stravithie, within the triangle formed by the Burghs. But in our corner of the constituency the chief damage was done to the Liberal cause by Asquith's security men, two detectives who accompanied him everywhere. 'If a man canna trust hissel' amang his [us], we canna trust him in Parliament.'

For among the fisher folk politics was largely a question of loyalties, and without mutual trust there can be no loyalty. Affairs of state might be beyond the comprehension of ordinary folk, but the principles on which true government rested were unchangeable and were the same as those which ruled the everyday life of the man-in-the-street, or should I say, the man-in-the-boat. My mother told me that when my grandfather realised how many millions had been added to the National Debt by the Boer War, he turned to my grandmother and said: 'Lizbeth, we'll never be able to hold up our heads again. That debt'll never be paid off.' Perhaps it is as well that he did not live to see the day when we would spend on armaments in one year of peace more than the total amount which made him hang his head in shame. Nevertheless, his spirit of independence and self-respect, seen and exercised on a national scale, might still have done something to restore the integrity which was once the hallmark of British politics.

I seem to have omitted any mention of the part played in politics, both national and local, by the press. This is a serious omission, for no account of life in the first twenty years of the twentieth century in the Coast Burghs would be complete without a record of our newspapers.

For many years there was one weekly paper published in Anstruther. This was the *Record*, owned and edited by Mr Louis Russell. Mr Russell was a prominent Liberal and so his paper was a Liberal paper. The dailies of course, published in Dundee, Edinburgh and Glasgow, were of every shade of political belief, and the readers chose according to their own views, but for the local news everyone was compelled to read the *Record*. Mr Russell had a brother, Mr C. S. Russell, who owned a printing works and did all, or nearly all, the local printing. He was a Tory and a member of the established church, while his brother, the Liberal, was a member of Chalmers Memorial UF Church. When political feelings were running particularly high, probably in 1910, it was almost inevitable that Charles Russell should bring out a rival newspaper, and so the *East Fife Observer* was born. For several years, Anstruther, with a population of about 3,000, supported two weekly newspapers, circulating solely in the Coast Burghs with a total population of about 20,000. Only political rivalry plus family jealousy could have kept up the struggle as long as it lasted, but the death of Louis Russell meant the death also of the *Record*, and the *East Fife Observer* remained until the 1960s, reporting faithfully the day-to-day news of the Coast Burghs.

10. Epilogue

I MUST FINISH AS I BEGAN, with religion, for since I started to write this book I have realised that while all the facts recorded are indeed memories of childhood, the interpretation of these facts owes much more to my last and closest association with the fisher folk, and this came about through religion.

As I have said earlier, I was converted as a schoolgirl and from then became what is known nowadays as a committed Christian. But my commitment was very superficial and formal until a serious illness in 1919 caused me to confront the fact that in all probability, medically speaking, my life would be a short one. This made me aware of spiritual values in a way hitherto unknown, and was in fact a preparation for all that was to follow.

I had for quite a long time been a member of a prayer group of the Faith Mission, quite a small group numbering fewer than ten at the end of the war. We met weekly in the home of an elderly man, a Baillie of the Burgh in Pittenweem, and he, his wife and sister-in-law were the 'elders', the rest being girls in their early twenties. In 1920 Baillie Toye moved to Aberdeen, and we were left without a leader and without a meeting-place. Had it not been for my new perception of spiritual things I think I would just have allowed the meeting to lapse, but somehow it seemed important that this little gathering should continue. We had no money to hire a hall, and we were all girls living in homes where there was no room which we could call our own, even for one hour in the week – not because of any unwillingness on the part of our families, but simply and solely because no one of us had or could have a room of our own. I was at that time in the Town Clerk's Office in Pittenweem, and was responsible, among many other things, for letting the town hall and the lower town hall. This latter was a somewhat grandiose name for a dark, dirty room with barred windows, which at one time had been the town jail, and had neither been altered nor improved since. But the rent was only one shilling and sixpence per night, and even if the others did not help, I could afford that much out of my salary of £120 per annum. So I booked the lower town hall for Thursday evenings and informed my friends.

It must have been in August that we held the first meeting, for it was still daylight, even in the dark ex-jail, and the children were playing in the streets. I think there were six of us, all dressed formally, for we were going to a prayer meeting and carrying our bibles. We made our way to the Wynd, and, as I turned the huge key in the lock, four little girls who had been skipping the open space before the town hall picked up their ropes and came over to watch. We disappeared within and closed the heavy door, but, nothing daunted, the girls pulled themselves up by the bars on the windows and so effectually shut out what light there was. I went out and pulled them down one by one, but by the time I had dealt with number four, number one was in position again. At last in desperation I said: 'If you want to see what we are doing come inside. You'll be less bother in than out', and in they trooped. They sat in a row on the front bench. Backless benches, one chair and a table comprised the furnishings. As we went through our usual routine of hymn-singing, bible-reading, prayer and address they were as quiet as the proverbial mice. I felt really sorry for them, prisoners of their own choice, in a cold dark room, while outside the sun still shone and other bairns were laughing and playing. So, when our service was over I said: 'You've been very good so I'll tell you a story all to yourselves', and I launched on my favourite pastime. They were still quite quiet but now it was not the quiet of boredom; they were entranced. When I finished and they reluctantly rose from their seats I said: 'We'll be back next week and if you come half an hour earlier I'll be here to tell you a story and then you can go home without sitting through the meeting.' Next week I was greeted by eight children, and week by week the numbers grew until I had over fifty, packed like herrings on the narrow benches.

Then the 'halflins' walked in. The 'South fishing' had started; all the men and older boys had left, and there was nothing to do in the dead town. So what better than to break up the bairns' meeting? I called on my friend the local policeman, and he arrived and turned them out; but I could hardly expect him to remain on duty from 6.30 p.m. to 9 p.m., and no sooner had he disappeared than they invaded the place once more. By changing my style of story-telling I managed to keep them in reasonable control for the childrens' meeting, but then they began to stay behind for the prayer meeting. Again we changed the pattern of our meeting and achieved some kind of discipline, but prayer of the kind we wanted was quite impossible. One evening, after I had told a gangster-type of story and preached what I thought was a gospel sermon, I appealed to their better nature. I told them how I had changed the children's meeting to suit them, how we had altered our prayer meeting into an evangelistic meeting on their behalf, and now, I said, we wanted to get down to the business which had brought us there, which was to pray. So would they please go home and leave us in peace? There were six lads of sixteen or so, hefty boys who could have lifted any of us with one hand. There was no chance of using physical force against them, nor could that have been our way. And then a remarkable thing happened. Week by week we had prayed for revival, and now it happened. The boys were held to their seats by a power outside themselves; they were 'under conviction of sin'. Our six lads went home 'converted'.

It was the autumn of 1920 and, as I have said, the boats were at the 'South fishing' in Yarmouth and Lowestoft, and there in East Anglia the last of the fishermen's revivals broke out. Many of our local fishermen were caught up in the movement and came north intent on spreading the fire.

The fishermen of Pittenweem returned home and immediately started evangelistic meetings every night in the Baptist Church hall, and the Cellardyke fishermen later followed suit with similar gatherings in the reading room of Cellardyke Town Hall. The local ministers were not consulted nor asked to assist, and at first were inclined to hold aloof. I am not sure how things developed in Pittenweem, since my first concern was for Cellardyke, but when the campaign was over, that is, when the winter herring fishing began and the men went back to sea, I was able to hand over my children's meeting to a local girl, a clerk in the post office, who carried it on for several years in more suitable premises in the Baptist Church schoolroom.

In Cellardyke the Faith Mission heard of the blessing that had come to the town, and two of their young men evangelists arrived to carry on the work when the fishermen had to give it up. For three months the meetings went on. As in Pittenweem the local ministers held aloof to begin with, but gradually they were drawn in and all of them were on the platform at the final meeting. Statistics mean little, but at the May Communion in Cellardyke I believe forty young men and women joined the Church and became the backbone of every department for the next thirty years or more.

When the campaign was over our Prayer Union group had grown from six to sixty, meeting weekly in the reading room. Again I felt a concern for the children whom I had been teaching in Sunday school for years. So I started a children's meeting for them, which met before the prayer meeting, and once again it grew by leaps and bounds until it was difficult to pack all the youngsters into the room. The reading room could hold about 80–100 adults, but on my register I had 200 children, with an average attendance of about 150, and mostly it was a one-man (or woman) show, for I had no regular helpers. That is how I came to know the fisher folk and their homes. I knew every child in the town, and when you know the bairns you soon get to know their mothers.

Of course my meetings did not go unchallenged, and I was soon in contact with the police, not by my choice but because of complaints from people living near to the town hall, who, like my aunt in my childhood, thought my hopeful band were young fiends. The local sergeant was a kindly man and he hated to interfere with what he thought was a very useful piece of work. He advised me to get the children to form a queue while they waited for me to arrive to open the doors and promised if they did so, nothing more would be said. I gave the children a straight talking to and they all promised faithfully that they would queue up 'like the Pictures' until I came.

Next Thursday I set out, without any misgivings, for my mile-long walk, after a hurried tea so that I should be punctual and not keep my flock waiting too long. Just as I turned from Rodger Street to Shore Street, Anstruther, I saw a whole procession coming to meet me; indeed as soon as they caught sight of me they started to shout: 'We've come to meet you.' When they clustered around, trying to hold on by a hand, an arm or the skirt of my coat, I asked why they had not made a queue as they had promised, and a chorus of voices answered: 'So we did, but we thought we would keep ourselves quiet by singing choruses, and the women came out and chased us away!'

It was during these last five years in East Fife that I really got to know the fisher folk. Whenever I appeared on the streets of Cellardyke I had a cluster of children round me, and if I went, as I sometimes did, from Caddies Burn to the Harbour Head I had a train of followers, almost rivalling the Pied Piper. The mothers and grannies soon got to know me; the older women recognised me as 'Alec Patrick's lassie' and the younger women remembered me as a playmate on the beach or hanging round the boatbuilders' yard, and it was as they talked with nostalgic regret of the old days that my childhood memories came into focus to form a complete picture in my mind.

By the 1920s the revolution (I merely mean this in its literal sense: 'the turn') was well on the way to completion. The young women had altered their way of life by the simple expedient of refusing to learn 'the mendin', i.e., the net mending. What they had never learned they could not be compelled to practise, so net mending, as the older women died or got past work, tended to become the responsibility of the agents who managed the boats. These employed local labour in most cases, but the family businesses of husband and wife, or sons and daughters as partners, had vanished.

I was accepted into the community in a way that few outsiders ever were. I even had a near-proposal of marriage from a young fisherman – not very bright I must admit – who began by giving me his idea of my good qualities in so far as they made me suitable for the high honour he had in mind for me. Just as I was wondering how to put him right in the kindest way possible, he came to the crux of the matter: 'But of course, there's just ae thing. Ye canna mend.' I agreed whole-heartedly that was an impossible barrier, and I nearly overdid it in my eagerness. 'You could mebbe learn, though,' he said hopefully, and I had to tell him that I did not even want to.

So now my story is ended, and I must leave my fisher folk. Even if it means that I shall be condemned as having become completely English, I must use one four-letter word which is, or was in my young days, completely barred in speech if not in writing, and leave them all my *love.*